THE SCULPTURE OF LIFE

THE SCULPTURE OF LIFE

ERNEST BOREK

Columbia University Press
NEW YORK AND LONDON   1973

Ernest Borek is Professor of Microbiology at the School of Medicine of the University of Colorado, in Denver.

Library of Congress Cataloging in Publication Data

Borek, Ernest, 1911–
   The sculpture of life
1. Cell differeniation. 2. Cell proliferation. 3. Cancer cells.   I. Title.
[DNLM: 1. Cell differentiation—Popular works.   2. Neoplasms—Popular works.
QH607 B731s 1973]
QH607.B67   574.8'761   73-6831
ISBN 0-231-03425-3

FOR SYLVIA

Preface

A pale winter sun shone on a heterogeneous group lined up in front of the southwest gate of the White House in Washington two days before Christmas, 1971. Among them were well-known Senators and Congressmen dressed in well-cut custom-made clothing; there were a couple of prominent ladies in expensive mink; and there were some less resplendent people dressed in the uniform of the working scientist, rumpled slacks and ill-fitting sports jacket. This unusual mix of people were all guests of the President of the United States. Guards in Graustarkian uniforms inspected the telegrams of invitation; demanded photograph-bearing identification; and waved guests in with a gesture which denoted more resignation than welcome.

Some hundred people, once cleared, were escorted into a large room half filled with television cameras operated by hirsute crews dressed with modish flamboyance: among the crew one could recognize English highwaymen, "Baddies" of our West, and Mexican bandidos.

The doors in the front of the room were theatrically flung open and the President of the United States was announced. The audience rose, the television cameras whirred, still photographers ran around to vantage points, and the President, obviously fresh from a television makeup technician's chair, turned on a contrived smile. He came to sign the Cancer Act of 1971. Cancer research had entered the political arena.

The Congressmen and Senators who guided the law into being smiled broadly as the cameras focused on them. Most of the scientists in the audience did not smile; many were worried. The hoopla surrounding the Cancer Act implied the conquest of cancer in the near future because a couple of hundred million dollars a year more were to be channeled into cancer research.

Those of us there who knew the "state of the art" had cause to worry.

Some of us had just returned from a series of conferences on what is *not* known about cancer, and what is yet to be done.

The Director of the National Cancer Institute, Dr. Carl Baker, had launched a unique program. He assembled some 250 of the leading biological scientists and cancer specialists, divided them into groups of 8 to 10, assigned each group to a well-defined area of the problem, and locked them up for three days in a resort outside of Washington.

Since this was probably the first time that working scientists were called in to assess current knowledge and to plan for the future, they responded and worked with a sustained intensity equaled only by the days before examinations for their Ph.D.'s or National Boards. The result is the National Cancer Plan, which is an inventory of research yet to be performed to understand and to control cancer. The list in fine print on both sides of 8 × 10 sheets is over a foot high.

That is why some of us there did not smile. The launching of the attack on cancer was compared by some to the program to land on the Moon. But, as one articulate scientist said, promising to conquer cancer by a broad assault now is comparable to trying to get to the Moon without knowing Newton's laws of gravitation.

Our disquiet stemmed from two sources. Promise of a cure within a defined time of any disease, especially something as complex as cancer, should never be made: There is no more cruel disappointment than promise of health not fulfilled. Moreover, what will be the effect on the attitude toward science and scientists of a public, already disenchanted with us, if we are copartners in making what amounts to a fraudulent promise.

Cancer is beginning to be understood, but no one who is well informed would put either a date or a price tag on its ultimate conquest.*

After the doleful ceremony, I decided to write a book to summarize for the intelligent reader what is known and not known about normal regulation and its derangements which result in the monstrous growths known collectively as cancer.

This is a report from the laboratory. It is an interim report on work that is just emerging, indeed, in some cases just starting. Some of these investigations may be false starts, but enough information is reasonably secure to convince us that we are witnessing the opening of a frontier which currently provides the greatest challenge and excitement for our intellect.

Growth and differentiation of a fertilized egg into the ultimate individual is so complex and so precise a sequence of molecular events that it borders on the miraculous.

The molecular biologist who would study differentiation at present must content himself with fragments of truth extracted with infinite patience and some ingenuity from the most complex creation in

* The price of redemption of promises of cancer cures has been mounting. In 1898, a Dr. Roswell Park conned the legislators of New York State to give him $10,000 a year and he would cure cancer. This was the beginning of the Roswell Park Memorial Institute in Buffalo, New York. Seventy-five years later this distinguished institution has a budget probably a thousand-fold higher than Dr. Park's original request and, while some first rate work has emerged from there, Dr. Park's promise, alas, remains unfulfilled.

the universe, a couple of hundred living cells performing their appointed tasks, ordered by infinitesimal molecular cues, to develop an individual. This process is equally awesome to contemplate be it the development of a starfish, a mouse, or a man. The miracle is not a specific species, or a particular individual; the miracle is the shaping of any life.

Fortunately, some intrepid experimental biologists have led the way into this labyrinth, and quite a few molecular biologists have followed. The motivations are varied; some are drawn to the challenge of the unopened frontier, some are awed by the problem of malignant growth and have decided to study not the abnormal growth but normal development. For, as the great French physiologist Claude Bernard, stated: "The abnormal is but an exaggeration of the normal."

At any rate, I feel that enough of a path has been made into this wondrously complex world by the tools and skills of molecular biology to invite the intelligent lay reader to share some of the excitement with the pioneers. In this connection it is a pleasure to report that the distinguished writer, Mr. Wallace Stegner, graciously consented to read this book in manuscript and has found it not unreadable.

ERNEST BOREK

Placer Valley, Colorado
January, 1973

Acknowledgments

Many friends have helped in shaping this book. It is a pleasure to testify once again to my indebtedness to Robert Tilley, Editor in Chief of Columbia University Press, whose encouragement helped to mount over my high threshold of doubt about the feasibility of presenting the subject to a general audience.

I am also indebted to several colleagues who, by reading some of the chapters, enlightened me and helped to keep errors, I hope, to a minimum: Cole Manes, Beatrice Mintz, Theodore Puck, Arthur Robinson, and Howard Ulfelder.

Specific acknowledgments are made on the appropriate pages to people who generously permitted reproductions of photographs or diagrams. However, I must give special thanks to my colleagues at this university, Keith Porter and Theodore Puck, for not only providing illustrations but also guiding me in their selection.

I am especially beholden to my secretary, Lynne Teller, for her ability to translate into readable script, tapes on which the barking of dogs and the chirping of the mountain birds of Colorado often all but obliterated the halting human voice.

E.B.

Contents

Science, like the arts, gives expression to the innermost yearnings of the human spirit and thereby enriches our lives.

SIR BRIAN FLOWERS

THE SCULPTURE OF LIFE

All our knowledge is ourselves to know.
ALEXANDER POPE

Chapter one

What Every Cell Knows

Every one of us is shaped by billions of cells. They are similar enough to form a harmonious whole but unique enough to form the different organs: heart, lungs, liver, kidney. If these billions of nucleated cells could be skillfully separated and knowingly manipulated, every one of them could give rise to a new man. This feat, which sounds like a science fiction fantasy, has already been achieved, to be sure not with human but with frog cells.

Dr. J. B. Gurdon, a biologist at Oxford University, is the man who demonstrated unequivocally that every cell with a nucleus, be it a cell in the skin or a cell in the lining of the intestine, carries the total information to make a complete, new individual. As so often happens in science, his telling achievement was built on groundwork prepared by others. Drs. Robert W. Briggs and

1

Thomas J. King of the Institute of Cancer Research in Philadelphia worked out with rare skill and patience the basic techniques which Gurdon employed so brilliantly.

The nucleus of an unfertilized amphibian egg is located just beneath the surface. Briggs and King made use of this anatomical happenstance for their manipulations. The nucleus can be pried out with a needle, or sucked out through an inserted microtubule; or the chromosomes in the nucleus can be destroyed by focusing a tiny beam of ultraviolet irradiation on them. Any of these manipulations leaves behind an essentially intact egg minus its nucleus. The next step is the insertion of a complete nucleus from any frog cell into the enucleated egg. Briggs and King hit upon a simple, ingenious technique for this transplantation into the foster egg. They used a tubule whose diameter was just too small to accommodate a frog cell but large enough for the passage of the nucleus. By sucking a cell into such a tube they destroyed the integrity of the cell but not of its nucleus. The latter could now be inserted into an enucleated egg and anxiously observed—with fingers crossed. A small percentage of such eggs with complete, living nuclear transplants would begin to divide just like a fertilized egg and would give rise to tadpoles.

Although these simple experiments, executed with expert skill, were impressive, their results could not be accepted as scientific proof of the presence of the total biological information for the individual in every nucleus. Devil's Advocates (and some scientists, especially those whose ideas are destroyed by a new, ingenious experiment, take on this role with inordinate zest) might argue as follows: Are we certain that the information for the growth of the tadpole came from the transplanted nucleus? Could there not be some residual information left in the enucleated egg?

Dr. Gurdon had the means to dispel any reasonable doubt about the validity of the conclusions. (Let us remember that the legal and scientific term "reasonable doubt" means a doubt entertained by a reasonable man. There are still a few scientists who reject these conclusions about the storage and utilization of biological information.) Gurdon had an indelible tag, a marker, on the transplanted nuclei which could be seen in the artificially created tadpoles. The nuclei of frog cells contain two distinct smaller bodies called the nucleoli. An observant graduate student at Oxford, Michail Fischberg, noted something strange in the cells of a line of South African frogs, *Xenopus laevis*. They had only one nucleolus! This is a genetic variation visible under an ordinary microscope which proved of immense value in answering a variety of questions.

Gurdon seized it to perform, among others, an impeccable experiment probing the amount of information in a nucleated body cell. He repeated the procedure of Briggs and King; he destroyed by ultraviolet irradiation the nuclei of unfertilized eggs of normal *Xenopus laevis*. (These have, of course, two nucleoli.) Then he removed cells from the lining of the intestine of the mutant strain of the tadpole, which has but one nucleolus. Cells of the lining of the intestine are highly specialized, endowed only with the molecular apparatus needed for their appointed task: to assimilate the components of the food as it passes within their proximity. Since they are in eternal darkness, these cells need no visual sensory molecules; they can make no hemoglobin; they need synthesize neither pigments nor claws. They are stripped down for the most effective performance of their specialized tasks. But do they carry any more information within them than is absolutely essential? Have they, as they were stripped for their specialized functions, shed the information for making claws, vocal chords, and visual pigments?

Gurdon was soon to know. He destroyed the nuclei of about 100 unfertilized eggs of normal *Xenopus laevis*. Into them he carefully inserted the nuclei from cells of the intestinal lining of the mutant frog with but one nucleolus. The frog eggs now contained nuclei endowed with the usual double dose of chromosomes. (The egg, of course, has only one half until the sperm delivers its quota during fertilization.)

Some molecular signal from the egg entered the nucleus and induced it to start yielding its hoard of precious information. A burst of molecular activity ensued which produced a doubling of vital components enabling the cell to divide, and the activity and cell division continued until a tadpole miraculously emerged. Only about 2 to 3 percent of such transplants were successful; but in these tadpoles the source of the information for their growth came unequivocally from the transplanted nuclei: Every cell of the semi-manmade tadpole contained but one nucleolus. In Figure 1.1 the reader can inspect a normal and a Gurdon-made *Xenopus laevis*.

Therefore every cell, however specialized and limited in its biological capacity, has the total information of the species tucked away in its nucleus.

What we know, and do not know, about the orderly—and, in cancer, disorderly—expression of all that biological information is the subject of this book.

Since it is within the cell that the controls over the expression of information operate, we must acquaint ourselves with its structure and molecular components. This will be a capsular summary of our knowledge of cellular structure and molecular function. If the reader is interested in the development of ideas that have been proposed, tested, rejected, or accepted to form our current concepts on biological information, he may wish to read my earlier book, *The Code of Life*.

One of the discoverers of the cellular structure of organisms, Schwann, described the total capacity of cells with profound insight: "The cells are *organisms* arranged in accordance with definite laws." The key word, of course, is organism. Schwann recognized that not only do cells form organisms but also that a cell is an organism unto itself. As we shall see in a later chapter, well over a hundred years after Schwann's categorical statement we were able to grow single mammalian cells in special media, enabling them to increase their components and inducing them to divide just as if they were some unicellular organism. The mammalian cell, therefore, retains the primordial capabilities of a unicellular organism, but through eons of evolution it has also acquired new knowledge: the capacity to perform a vast spectrum of novel roles.

In Figure 1.2 is an idealized plant cell as seen under an electron

1.1 A & B Tadpoles Developed Normally and by Nuclear Transplantation

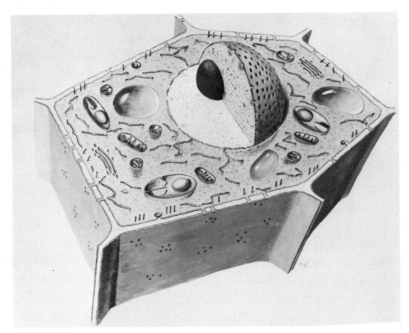

M. C. LEDBETTER AND KEITH R. PORTER

Figure 1.2. Typical Plant Cell

microscope via the searching beams of electrons. Let us look at the structures of such a cell in detail. The cells of plants and of bacteria are enclosed by semirigid cell walls, but animal cells are not endowed with this armor.

All cells, whether of plant, animal, or bacterial origin, are completely enclosed by a membrane which guards the physical integrity of the cell, maintaining it as an inviolate, semiautonomous unit. Not only does the membrane guard the cell by physical containment, but it also mounts an unceasing guard over the ports of entry on the frontiers of the cell. The membrane recognizes with its uncanny molecular memory the hundreds of compounds swimming around it and permits or denies passage according to the cell's requirements. Unknown compounds which fail some subtle test for recognition on the molecular ramparts are usually excluded, and the cell is thus guarded against their possibly harmful presence.

The cell's membrane girds a kingdom of varied topography. Some of the structures, such as the organelles, which must be present in every cell of a given tissue, are obligatory. There are other structures whose distribution is more random in otherwise homogeneous cells. These are the inclusions, which may be a tiny blob of fat or a grain of starch. Organelles must be reproduced during cell division; inclusions need not be. The function of some of the organelles in the cell is well known; others still manage to hide their activities.

Our knowledge of the function of the organelles depends on our ability to remove them in more or less intact form from the cells. In order to achieve this the integrity of the cell must be destroyed. The destruction may be achieved by mechanical grinding with some abrasive powder; or it may be done by exposure of the cells to sonic vibration; or, in the case of bacteria, the cell wall can be dissolved by an enzyme, and without its protective enclosure the cell pops and disintegrates like an overblown balloon.

The next step involves a series of centrifugations at higher and higher speeds and, consequently, at gravitational forces of increasing intensity. Exposing the cells' contents to 100,000 times the force of gravity is a routine operation in most biochemical laboratories today. The various components of the cell sediment out sequentially as increasing gravitational forces are imposed upon them. The many different fractions are harvested individually; their appearance is examined under an electron microscope; visual correlations are then attempted with the organelles as they appear in the intact cell.

With luck we can obtain preparations of some organelles that retain some of their original functions. For example, there is a widely distributed structure which is just barely visible under an ordinary microscope: the mitochondrion. The reader can see a photomicrograph of a mitochondrion in Figure 1.3. We can get an approximation of their real size if we attempt to visualize a thousand such mitochondria forming a ladder across a dot on a letter "i." Yet this tiny component of the cell has an elaborate structure, which is emphasized in the idealized drawing of Figure 1.4.

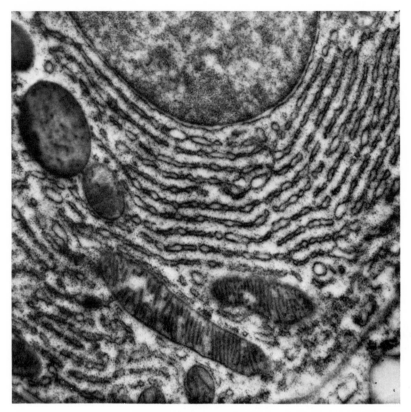

JOURNAL OF BIOPHYSICAL AND BIOCHEMICAL CYTOLOGY

Figure 1.3. Mitochondria

The function of the mitochondrion emerged in the early 1950s as a result of two entirely different lines of investigation. On the one hand, there were investigators interested in the electron microscopic anatomy of the cell who were concentrating the mitochondria by selective centrifugations. About the same time, biochemists who were interested in the mechanisms with which the cell generates its energy from glucose were concentrating particles from disintegrated cells that could carry on oxidation even outside the cell. When the two groups, the anatomists and biochemists, compared their preparations they found them to be the same: The mitochondrion

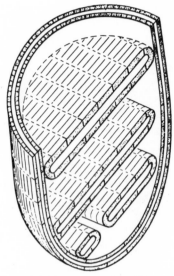

Figure 1.4. The Mitochondrion Dissected

turned out to be the furnace of the cell. It is a furnace with at least forty different working enzymes embedded in it in fixed positions. The enzymes, like so many workers on an assembly line, oxidize glucose and deposit its precious energy into a form readily usable by the cell for a multitude of tasks that require energy.

Another organelle whose function is known is the chloroplast, which is found in plant cells and in some bacteria (see Figure 1.5). The function of chloroplasts was relatively easily deduced. The telltale green color of the preparations indicated the presence of chlorophyll, the catalyst of photosynthesis. Thus chloroplasts are the structures on which all forms of life eventually depend. They alone have the life-generating ability to concentrate the sun's prodigious but diffuse energy into a form which can be the fuel of life. With extraordinary efficiency they harness the electromagnetic energy of light and use it to pack random, disorganized molecules of carbon dioxide into the highly organized, energy-rich structure of the glucose molecule. In turn, the enzymes in the cells of every living organism can reverse the process and tap the energy within the glucose molecule for their own needs as they dismantle it to carbon dioxide.

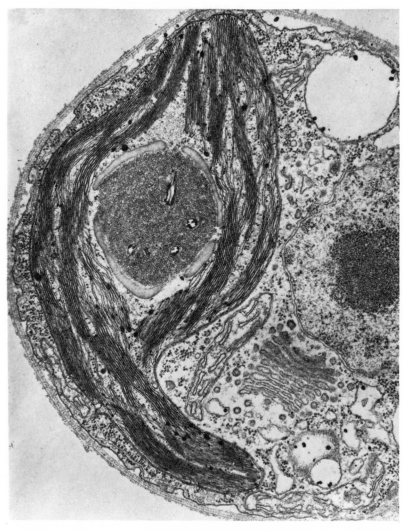

G. E. PALADE

Figure 1.5. The Chloroplast

Living organisms can perform various transformations of energy. The ear converts sound and the eye converts light into electrical energy. The skin can translate mechanical pressure into an electrical signal, and the nerve can transform chemical into electrical energy. Some organisms such as the firefly can use their stored chemical energy to generate light. Our muscles consume chemical energy for motion; our vocal chords do the same but produce an exquisitely controlled motion which gives rise to sound. The source of energy for all these transformations is the specialized molecular battery, adenosine triphosphate (ATP), which is produced in the mitochondrion from the chemical energy of the glucose molecule. In turn, it was the chloroplast which, with virtuoso skill, trapped the energy of light and transformed it into the energy that holds the glucose molecule together.

Figure 1.6 is a remarkable microphotograph of an algal cell in which the product of the chloroplast, starch granules (the white popcorn shapes), can be seen emerging.

Although we have been studying the cell intensively for 130 years, we can still discover new structures in that wondrously complex speck of life. The lysosome is a recently identified unit; it approximates the mitochondrion in size but lacks its highly organized structure. The lysosomes seem to be bags of digestive enzymes that can dismantle large protein and nucleic acid molecules. The rubble of small fragments so produced can pass through the membranes of the mitochondrion and be consumed in its furnace.

It is obvious why such potent agents within the cell must be contained; otherwise the contents of the lysosomes would destroy the cell by literally boring from within. How the structures that are doomed for dismemberment pass into the lysosomes is obscure at present. This may be the method of elimination of defunct or excess structural or functional components of the cell. Studies with isotopic tracers have shown that the molecular components of all living things are in a constant state of flux; tissues are being built up and broken down simultaneously. The molecules which compose our bodies today will be gone in a few months and will be replaced by new ones from our foods. Even highly organized structures such as the mitochondria have but a transient life within the cell. The life span of a mitochondrion of the liver has been estimated to be from 10 to 20 days.

The reason for this profligate discarding of cellular components is now well understood. It lies in the mode of storage of biological information. Let us assume a theoretical biological molecule, say

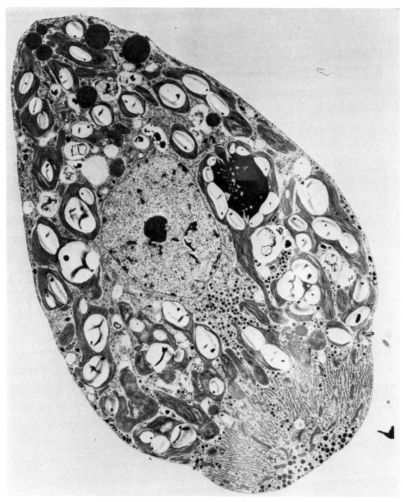

Figure 1.6. Chloroplasts at Work

a protein, with 100 linear component units. The fidelity of its sequence of components must be absolute; otherwise it would lose its specific attributes. The information for the synthesis of this sequence of 100 components is stored in the genetic material as a single continuous entity consisting of 300 components. (The reason for this will be described later in this chapter.) A protein of 100 component units can be singly cleaved in 99 different places. (The permutations of multiple cleavages is beyond my mathematical ability to compute.) Specific information for the repair of 99 cleavages with complete fidelity to the original sequence would be beyond the ability of the cell to store.

It is easier to assemble a brand new mitochondrion than to repair one whose structure has developed some molecular fissure or whose functional efficiency has faltered. An uncannily perceptive method of recognition must single out the defective structure for dismemberment by the enzymes of the lysosomes, and the resulting structural debris is fed in as fuel to the furnace of intact mitochondria.

The Golgi bodies, which were discovered in 1898 by the Italian whose name they bear, are still another example of well-defined organelles. However, despite their venerable history and ubiquitous distribution in animal cells, very little is known of their function. One hypothesis is that the Golgi bodies are the sites where some of the proteins are assembled and packaged for export out of the cell. However, this must remain as an unverified conjecture until we can isolate these structures in concentrated form and induce them to perform their tasks—in the jargon of science—*in vitro*.

Very recently a heretofore unobserved organelle, the microtubule, was discovered by Dr. Keith Porter, who seems to be equally endowed with extraordinary technical skill in electron microscopy and exceptional pictorial imagination. Many electron microscopists have observed what appear to be tiny holes when a cell is sliced, removing its lid, as it were. But Porter can slice cells thinner and observe better than anyone else. In the next slice of the same cell he could observe another hole which was directly beneath the hole in the upper slice. Porter recognized that he was slicing a tube crosswise. After Porter's discovery, others developed methods of isolating intact microtubules and still others studied their method of synthesis. From these studies emerged one more sophisticated expertise of a living cell: It assembles long tubules just the way a plumber assembles them from short pieces fitted together with appropriate coupling devices. However, the analogy with plumbing ceases there. The microtubules do not seem to be conduits for fluids; their tubular shape appears to be designed to afford maximum strength with minimum weight.

What other organelles may be in the cytoplasm but too small for recognition with our present methods of observation, we do not know. Faultless structure is the hallmark of the cell, and we can, therefore, predict that in the range between single molecules and relatively giant structures such as the visible organelles there must be intermediate organizations designed for some special tasks. The detection of these ''microorganelles'' will be the task of future cytologists.

Now let us come to the site of the storage of biological information, the nucleus. The nucleus is a spherical or ovoid structure encased within a membrane. An idealized version of it can be seen in the center of Figure 1.2. Usually there is only one nucleus within a cell; however, some cells are endowed with two or more. On the other hand, not every cell possesses a well-defined nucleus. For example, whether bacterial cells contain a genuine nucleus was a controversial topic among cytologists until recently. The electron microscope resolved the argument, and it is now generally conceded that bacteria are devoid of this structure. (The bacteria do, however, have the usual genetic apparatus, although it is not packaged in a nucleus.)

During part of its lifetime the nucleus appears to be a homogeneous structure except for a small sphere within the sphere, the nucleolus, which stands out as a discrete area. The nucleus sometimes gives the impression of unchanging serenity. This period in the life of a cell was therefore called by early cytologists the ''resting'' stage. The description is valid only so far as visible changes are concerned. Beyond the reach of even the most powerful electron microscope there is a maelstrom of activity going on within a resting cell with explosive speed and boundless variety. The most elaborate factory with the longest assembly line is a toy compared to the cell's machinery during the ''resting'' stage. Every component of a living cell is being manufactured at this time at a rate sufficient to ensure that, when the time for division arrives, there will be a full stock to draw on to shape the many structures needed by the two daughter cells.

During the resting state, also called the interphase, the chromosomes are diffuse and give the appearance of aimless disorganization. However, the appearance is misleading, for analysis reveals that the chromosomal material is increasing to ensure that there will be an adequate dowry of this precious substance for each of the daughter cells. At some point during the accumulation of the cell's substance, stock is taken, a decision to divide is reached, and the first steps to carry out the decision begin. Two pairs of minute structures—the centrioles—which lie just outside the nucleus begin to move apart, and they stretch between them gossamer threads or spindles which are microtubules. By the time the centrioles reach the opposite ends of the cell the chromosomes become more coiled and condensed. This is the beginning of the so-called prophase.

Then the membrane which up to now encased the nucleus begins to crack and crumbles like the walls of Jericho. The potent molecular trumpets in this case must be some enzymes, released perhaps from some lysosome. The thin threads of the spindle can now reach across the whole cell, unhindered by the barrier of the nuclear membrane. By some mechanism still unknown to us the threads of the spindle become enmeshed with small bodies within the chromosome. (Their position in the approximate center of the chromosome gives them their name, centromeres.)

Whether the entaglement of the centromere and spindle thread is achieved by some molecular hook reaching out or whether the centromeres themselves dole out their threads to meet the spindles is unknown. Whatever the lassoing mechanism, the chromosomes are now secured to the centrioles like an incoming ship tied by a tossed mooring line to a cleat on the wharf. The chromosomes, which are now well-defined, rod-shaped structures, are dragged into the "equator" of the cell and are lined up, equally divided, like sausages carefully doled out by a judicious parent between two equally deserving children. This is the metaphase. A movement toward the opposite poles slowly begins. And now the profound wisdom underlying all of these convolutions and alignments becomes obvious. The chromosome is a long, frizzly coil. When fully extended, the threads of the human chromosomes could be a meter long. During the prophase stage of mitosis this tangled thread is wound into a twin set of 46 chromosomes, each but a few millionths of a meter long. Like so much fuzzy unmanageable wool, the stuff is carded and wound into small tight spools.

Then during the movement of the anaphase the chromosomes can break cleanly without danger of entanglements of the coils, assuring a clean separation and even distribution of the chromosomal material. The dragging of the chromosomes now can begin. That it is really a drag, the spindle fiber being taken up like a cable by a capstan, is obvious from the shapes the chromosomes assume during their voyage. They look like a piece of rope snared by a fisherman's hook: The point of attachment swims ahead of the rest. Depending upon the location of the attachment of the spindle fiber—in other words, the position of the centromere—the chromosomes often assume characteristic shapes during their migration toward the poles. Some assume the shape of a "v" others of a "j", but this hydrodynamic distortion is by no means the only influence which determines the shapes of chromosomes. Their sizes and shapes even at rest are varied. The distance traveled by the chromosomes can be considerable, as much as five or ten times their own length. Their speed, however, is less than reckless—about a millionth of a meter in a minute.

At the end of their journey (the telophase) the chromosomes, as if tired of the long period of confinement, uncoil and stretch. But, lest they reach into the cytoplasm and thus wreak havoc with the cell's organization, new nuclear membranes are spun around them. The cell walls or membranes begin to be pinched off in the center and soon there are two young daughter cells, complete in every way, ready to start a life of their own. If this mitotic division is repeated enough times to yield 10^{14} cells with proper differentiation so that the appropriate specialized cells are produced, then we have a whole new organism, a wondrous creature, such as the reader of this book.

An idealized, diagrammatic representation of various phases of cell division in a plant cell can be seen in Figure 1.7. In these excellent three dimensional projections, the dominant role of the microtubules in the division process, may be clearly seen.

One of the pragmatic tests a hypothesis must pass in order to gain validity is the prediction of the existence of new phenomena. This can be done with high frequency in the physical sciences,

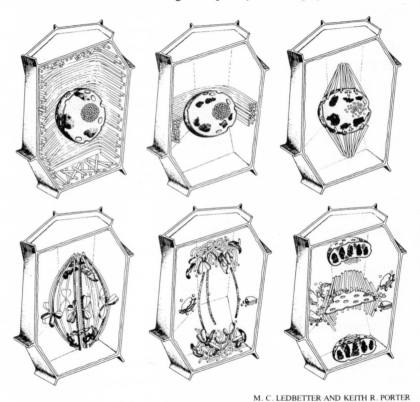

M. C. LEDBETTER AND KEITH R. PORTER

Figure 1.7. Cell Division in a Plant

for inasmuch as God, who created our physical universe, was a mathematician, there is a regularity, an order, in the universe. But the God who created the living world was a pragmatic empiricist. We are here, as we shall point out during the course of this book, as a result of an uncountable number of mutations. A living organism of today is an integrated product of all of those chance alterations.

But, in the midst of this welter of change, some basic attributes of structure and patterns of function endured. No living creature that lost its nucleic acids or proteins survived, no organism exists without the battery of life, ATP. Even the bacteriophage, this simplest speck striving to be counted among the living, contains nucleic acid, protein, and ATP, although by itself it is unable to make any of these prime structures of life.

In living functions, too, there must be some mechanisms so fundamental that no organism can survive their derangement. The orderly apportioning of the chromosomes into the sex cells is such a basic mechanism that it could have been predicted—provided one had had the insight of a genius. August Weismann, a German biologist, had such insight. He considered what might happen if each sex cell, the male and the female, contained the full complement of the chromosomes of the parent. In each fertilization, therefore in each generation, the chromosome number of the offspring would have to double. In not too many generations a cell would be as crowded with chromosomes as a football stadium and would have to be as big to accommodate all of them. Weismann, who believed that hereditary traits are entrusted to the germ plasm and rejected the theory of inheritance of acquired characteristics, could not countenance such chaos in the cell. He made one of the few bold predictions in the history of biology. The number of chromosomes must be reduced to one-half in every sex cell.

Patient work by several biologists in the last decade of the nineteenth century proved the soundness of Weismann's predictions. During the shaping of the germinal cells, both male and female, a reduction in chromosomes to half the total number in the parent's body cells occurs. This kind of cell division is called meiosis, from the Greek "to reduce." The diminution of chromosomes to one-half in the "haploid" sex cells can be seen under a microscope during meiosis, and, after chemical analysis of the stuff of the chromosome became perfected, it could be demonstrated by chemical measurement as well.

It is within the chromosomes of each sex cell, then, that the information and the promise of a new living creature is enshrined. But the chromosome is a highly complex molecular structure: It contains proteins, nucleic acids, some fats, and some sugars. Is the information for the creation of a new individual—indeed, for the perpetuation of the species—inscribed in one of these molecular components, or in all of them? Our willingness to propose such a question implies a leaning to a reductionist philosophy: Life can be interpreted on a rational basis within the framework of the laws of the physical sciences. Pillars of confidence in such a philosophy have been erected by experimental biologists of the past few decades.

As recently as 50 years ago an entirely different view, a holistic mysticism, prevailed. Witness a statement by a prominent English biologist, William Bateson, in 1916: "It is inconceivable that particles of chromatin or of any other substance, however complex, can possess those powers which must be assigned in our factors (genes). The supposition that particles of chromatin, indistinguishable from each other and almost homogeneous under any known test, can by their material nature confer all the properties of life surpasses the range of even the most convinced materialism."

The holistic philosophy of Bateson, and he has a few spiritual descendants today, is essentially one of desperation: If hereditary information is the summation of all components of the individual, we should not even attempt to study its molecular mechanism. Molecular mechanism can be meaningfully studied only in relatively pure systems, and what good is it to purify one of an almost infinite number of biological factors contributing to the total heredity?

Fortunately not all biological scientists were shackled by defeatist presumptions. Oswald T. Avery, a research physician at the Rockefeller Institute, believed in the biochemical approach to the study of the attributes of purified components extracted from living organisms. His faith, insight, and experimental skill revealed to us the mechanism of storage of hereditary information. Avery followed up an observation of an English experimental pathologist, Griffith, who had stumbled on an extraordinary phenomenon. Griffith injected into mice live, innocuous pneumococci along with steam-killed virulent pneumococci. The mice that received both preparations died of pneumonia, and on autopsy it was found that they were teeming with live, virulent pneumococci. Moreover, the living pneumococci bore stigmata of the virulent organisms; they were endowed with a characteristic glistening coat which is absent from the non-pathogenic ones.

The singular discovery was rooted in one of two mechanisms; either the dead, steam-killed bacteria were somehow revived, or an equally implausible, almost supernatural event occurred: The innocuous bacteria without the telltale glistening coat were *transformed* into organisms endowed with both attributes, lethality and the identifying coat. Griffith lacked the biochemical skills to explore the startling phenomenon; but Avery who, although he was also a physician, mastered the tools and principles of biochemistry, repeated and had confirmed Griffith's findings and also recognized their momentous meaning. A strain of bacteria can be endowed by the presence of another strain of dead bacteria with the biological attributes of the dead organisms.

Avery went beyond Griffith's experiments; he could take *extracts* of the dead virulent bacteria and add them to the nonvirulent organisms and convert them into virulent ones. Since these transformed bacteria, as they came to be known, would continue to reproduce and perpetuate their newly acquired attributes, they had undergone a hereditary alteration. Avery recognized that he had accomplished nothing less than an intrusion into the genes of bacteria, for it is in the genes that hereditary traits are enshrined.

A scientist on the frontier of knowledge is like an explorer who must make a series of decisions. Do I go on? Is there any promise or hope that further exploration may be fruitful? If he makes the wrong decision the explorer may risk his life; the scientist risks his reputation and, lately, the emoluments of a successful career.

Avery, intrepid explorer that he was, was not frightened by the awesome, implausible task he set for himself: the exploration of the nature of the gene itself.

Once a major decision is made, minor ones present themselves.

If I go on, which way do I go? The directions are pointed by a vector made up of intuition and personal prejudice. Those who would have dared to suggest that the functional part of the chromosome is a single, identifiable entity would most likely have bet on its protein component as having the central role. The voice of logic dictated such a conclusion. Proteins are made up of chains with perhaps a hundred links of 20 different amino acid components. The permutations of variability are astronomical: enough to accommodate the multitude of known heritable traits.

Avery's own prejudice led him to favor some complex sugar as the genetic material, because of his chemical knowledge that the shiny coat which was an attribute of the transformed pneumococci was composed of complex sugars.* In the early 1940s, it is safe to say, few if any scientists, among them Avery, would have put their money on another cardinal component of the chromosome, deoxyribonucleic acid, better known by its acronym, DNA, as the functional component of the gene. Though well-known and chemically well-characterized for decades, DNA was brushed aside as an unlikely candidate for a pivotal role in storing genetic information. This rejection was again dictated by the voice of logic: DNA's structure is too simple for the infinite capacity demanded of a reservoir of genetic information.

DNA, though vast in size, is very constrained in its component parts. A sugar (deoxyribose), phosphoric acid, and only four other components, the nitrogenous bases (adenine, thymine, cytosine, and guanine), compose DNA, and in the early 1940s it was believed that even these possible sources of variability were equally distributed in different DNAs. Moreover, the intellectual spirit of the time was preoccupied with catalytic attributes and, in these, proteins are the performers without peer: Some proteins can perform feats of chemical sleight of hand a million times a minute. No one could show any catalytic attribute of DNA. Indeed, the most educated guess about its function was that it was a reservoir of

*As it turned out, the shiny coat contributed to virulence via a trivial mechanism. It protects the pneumococci against dissolution by the enzymes of the infected host.

its components, such as phosphates, ready to release them as needed. At most it was thought by some that DNA is a supportive matrix on which the truly functional genetic proteins are stretched.

Avery patiently studied the extracts of pneumococci with their gene-changing potency. He isolated the known components—proteins, sugars—with the standard methods of chemistry. They were totally impotent in transferring virulence to the innocuous pneumococci. But the residue in the original extracts was still active. Analysis of the residual component with the biological potency revealed it to be DNA.

Molecular evolution had a surprise in wait for us. DNA is inert and by itself it is as informative as the latest encyclopedia in the hands of Neanderthal man. DNA, to express itself, needs a system of translation involving literally dozens of proteins and three other nucleic acids. No one could have predicted this.

The importance of this flash of revelation was not lost on Avery, a man of boundless, but disciplined, imagination. However, its impact on the whole scientific world was that of a pebble in the ocean. The reason for this was twofold. Avery published his paper in 1944, when many scientists had much more compelling preoccupations than current advances in science: The world had to be secured free for science and other creative activities; indeed, for humanity itself. Moreover, communication in science was very limited. Basic science was still practiced as a hobby of a few very talented individuals. The amount of support from society or from governments, except for science with practical ends, was minimal. Avery's communication appeared in a medically oriented journal, and there were no vehicles as there are today which give notices of all articles in all journals. Today it is relatively easy to find out what was published last month in the *Journal of the Bulgarian Academy of Sciences*. As an example of how slowly information about undoubtedly the greatest single biological experiment of the century spread we might note that Avery, though he lived for 10 years after his discovery, which was to re-orient all of our thinking, did not receive the Nobel prize.

But slowly the flash of Avery's genius reached the eyes of other biological scientists, and it quickly became apparent that the new information posed a conundrum. The dynamic functional parts of all living cells are the proteins. These are the components with catalytic properties which carry on the multitude of functions of the living cell. Individual proteins are composed of different permutations of the same 20 amino acids. Thus for their precise synthesis there must be information for the alignment of 20 different component amino acids. If these 20 amino acids are to be specified by some structure in DNA, there must be at least 20 different and variable entities in DNA itself. However, DNA, as we stated earlier, has only four variable components, the four bases—adenine, cytosine, guanine, and thymine. How can a four-component message system be expanded to encompass 20 different signals?

The problem is the same as that faced by the inventor of the Morse code. How can a telegraph which has only 2 message components, a dot and a dash, encompass the 26 letters of the alphabet? The riddle was solved by two scientists almost simultaneously. Since there are only four variables in DNA and they must be stretched to the 20 different components of proteins, it can be done only one way. More than one component of DNA must spell out an amino acid, and thus the number of possible signal units in DNA can be extended. If 2 of the DNA components stand for one amino acid, then the number of possible permutations available within DNA is 4^2 or 16. This is, of course, inadequate, falling short by 4 of the required 20. However, if 3 of these base components of DNA are the signals for one amino acid, then the number of possible permutations of variability is increased to $4^3 = 64$, which is more than enough to accommodate all the known amino acids.

This is very primitive numerology. It was almost an insult that molecular evolution could not devise a different, perhaps more complex, system of signals. These could be the geometrical configuration of the various bases or combinations of them. However, 25 years after these primitive suggestions were made, we can state with confidence that this simple numerology is indeed the basis of the storage of genetic information. Three bases in DNA represent a single amino acid of a protein. This confirmation was gained through the efforts of scores of ingenious biological scientists who came to be called molecular biologists.

I do not have space to describe the whole history of these exciting discoveries. I may again refer the reader to my earlier book. I shall only give the conclusions as we know them today, because these in turn will be the concepts by means of which we can understand the rest of the book. For it is the belief of most of us that the sculpture of life is shaped by these molecular mechanisms as we understand them today.

DNA is made up of a long strand, held together by molecular forces, whose prominent and variable feature is the sequence of the four bases. Eight years after the discovery of DNA as the genetic material, it was deduced by Watson and Crick that DNA is a duplex strand, one strand being essentially a mirror image of the other one. But the images are expressed in molecular structures. Thus an adenine is paired or opposed by another base, thymine, in the second strand, and a guanine in the first strand is opposed by a cytosine in the other one, and there is an invariable tendency for these two pairs of bases to align just that way.

The shaping of DNA this way is another of the remarkable accomplishments of molecular evolution. For by this simple mechanism two things are achieved. The DNA itself is strengthened: If there is a stress on one of the strands cleaving some of the chemical bonds, the other one, the mirror image strand, which adheres, will support it until repair mechanisms arrive. But, more importantly, this mirror image structure assures the almòst infallible duplication of the information entrusted to DNA as the cell divides. Each strand can serve as a mold on which the second strand can be shaped, thus assuring equal distribution of the information of the DNA and also assuring continuity. DNA in each of us today is a direct, continuing descendant of DNAs which existed millions of years ago.

It is quite obvious that the DNAs of various organisms must be very similar. This is inevitable since most of the structure of all DNAs is the same. It is only permutations of the four bases which are variable. How could the species individuality of DNAs be ensured? The maintenance of such individuality is a categorical imperative; otherwise, the DNA from invading parasites could be easily integrated into the DNA of the host by a whole battery of DNA-sealing enzymes. It would be a grievous calamity, if, say, the DNA of some flagellar protozoan became integrated into a human DNA. The offspring of the hapless lady might be mermaids.

Obviously, a safety device had to be invented. The device is a group of exquisitely knowledgeable enzymes which, like branding irons, stamp a species individuality on all DNAs. The brands, however, are the same. They are small atomic configurations known as methyl groups, and their frequency and position in the DNA have a species specificity.

The assumption that DNA is a structure which can be easily duplicated so that its integrity is preserved and handed down almost indefinitely was confirmed by a discovery of the American biochemist, Arthur Kornberg. He found an enzyme which is capable of reconstructing DNA provided that a model, a template of DNA, is present.

These findings, exciting though they were, still told us nothing about the mechanism of translation of the code of DNA into a sequence of amino acids which make the proteins. But this information, too, came very soon. It had been noticed by workers that whenever there is protein synthesis, there is within the cell also synthesis of another nucleic acid, ribonucleic acid. This too is a very large molecule which differs from DNA in two respects. Its sugar component is different; it is ribose instead of deoxyribose. This is the component which confers the name on this cellular component: RNA. Still another difference between RNA and DNA is in the configuration of one of its four component bases; instead of the thymine of DNA there is uracil in RNA. The other three bases are the same.

The most informative work on the detailed mechanism of the synthesis of a protein came from the work of a research physician biochemist, Dr. Paul Zamecnik, of Harvard. He and his associates achieved protein synthesis in a test tube for the first time with an unknown mixture of cellular components. However, by a process of elimination they were able to show which components present in their cell mixture are really essential for protein synthesis. Among these are large structures called ribosomes which contain both RNA and proteins, a smaller structure called transfer RNA, and an energy source in the form of the cell's battery, ATP. They missed one type of RNA, but this was soon to be revealed by the work of others. It is known as messenger RNA. This very complex system has been beautifully conceptualized by two Frenchmen, Dr. François Jacob and Dr. J. Monod of the Pasteur Institute. We shall use their nomenclature in describing the process of recovering the information encoded in DNA.

The DNA of the nucleus can not only be replicated to make more of itself; it can also be copied into a new macromolecule which then can be transported out into the cytoplasm where protein synthesis largely occurs. The copying of DNA is called transcription. The sequence of bases containing the information in DNA is copied by an enzyme which makes a strand of RNA that is the mirror image of the structure of the original DNA. In other words, if there is a base, guanine, in DNA, there will be a base, cytosine, in the newly made RNA. The name messenger RNA has been given to this information-bearing molecule.

Messenger RNA carries its information to the ribosome, where a new process, translation, ensues. Transfer RNAs are waiting at the ribosomes. They are essentially RNA, but they have the most complex structure of any of the macromolecules in a cell. They must have such a complex structure because they have many functions. Transfer RNA must be loaded with a particular amino acid and, without fail, it must be loaded with only one specific amino acid. Then transfer RNA is attached to the messenger RNA by a system the crux of which is the recognition of the three bases coming from DNA which specify that particular transfer RNA, with its particular amino acid. Thus transfer RNA is the most knowledgeable of molecules. It must know the DNA and

RNA code and it must also somehow recognize the appropriate amino acid which is assigned to it. Other enzymes now complete the linking of the amino acids which have been aligned in appropriate sequence, forming the protein chain.

These were brilliant contributions which revealed the overall machinery of the extraction of information from DNA. However, all this still told us nothing about the actual sequence of bases which make up the code for each amino acid. At that time we were in the position of someone who heard signals, dots and dashes, coming in over a wireless in a more or less repetitive pattern but did not know the Morse code and so was unable to decipher the message. The Morse code of the living cell was decoded by a young man, Dr. Marshall Nirenberg. As so often happens in science, this was not his original goal. Being a perfectionist, he cleaned up the system of translation so well that he could detect a chance, faint message, and thus it was he who broke the genetic code.

Dr. Nirenberg used essentially an artifically prepared messenger RNA which contained but one of the four bases. He had a messenger RNA with nothing but a long sequence of one of the bases, uracil. When he added this synthetic messenger RNA to his protein-synthesizing system, to his great surprise only one amino acid, phenylalanine, was used to weave an artificial protein molecule. What he obtained was a long sequence of phenylalanines attached to each other. Dr. Nirenberg immediately recognized the meaning of his finding. He had a repeated signal and thus the signal continued like a broken record repeating, in this case, the code for phenylalanine. Therefore, assuming a three-base code, three uracils signify a phenylalanine.

Once the clue to the decoding of the first letter of the genetic alphabet was available, the rest was easy. It served as the Rosetta stone for deciphering all. Other synthetic messenger RNAs could be made by permutations of the four bases into the 64 triplet possibilities. Soon all the amino acids could be assigned an appropriate genetic code in the messenger RNA and, in turn, in the DNA.

To our surprise, for most amino acids there was more than one code. Again let us recall that there are 64 possibilities of permutations for the triplet code from the four bases and in turn there are only 20 amino acids. The significance of this multiplicity of the code did not become clear until later. Some of these codes signify that the amino acid is to be in a certain position of the protein chain. There were two triplet codes to which no amino acid could be assigned and, confirming the validity of the whole interpretation of this mechanism, we found that these orphan codes which have no amino acids assigned to them merely serve as punctuation. They do not represent any amino acid and, therefore, when they appear in the chain their presence indicates that the message is terminated, since no amino acid can be inserted to break the gap between the last amino acid and any subsequent amino acid.

There are several other lines of evidence which confirm the validity of this triplet code. The most compelling one comes from the work of a brilliant biochemist, Dr. G. Khorana, who was working at the University of Wisconsin. He synthesized a synthetic messenger RNA which contained two bases in alternating sequence: ABABAB. If the triplet code has any validity, then this synthetic messenger RNA with the repeating two components could spell out two different code words, namely, ABA and BAB. Khorana chose the components A and B carefully so that ABA was the code for one amino acid and BAB for another one. If the triplet scheme is valid, then in a protein-synthesizing system this synthetic messenger RNA should produce a protein containing only two amino acids in alternating sequence; and that is indeed what he found.

This was the crowning achievement in a whole series of brilliant experiments starting with Avery's discovery. It is a tribute to the capacity of the human mind and to the persistence of the biochemists and the molecular biologists who were not shackled by holistic mysticism and were therefore ready to explore the unexplorable. But we must reserve our awe for the extraordinary feats of molecular evolution which through eons of trial and error evolved this remarkable system. "Evolve" is possibly the wrong term. "Created" is more appropriate, and I use the term creation in its Biblical meaning. The genetic code had to be created anew. It has no counterpart in Nature. In the nonliving world there is not even the crudest counterpart or the most primitive forerunner of the storage of information. Catalysis by proteins has numerous facsimiles in organic and inorganic chemistry, but the specifying of a particular amino acid for protein synthesis by a code of three bases was an absolute invention: It was an awesome act of genius.

But, as far as we know, all of the information in DNA and consequently in RNA is linear information. All that evolution managed to accomplish in the storage of information is the synthesis of a linear sequence of amino acids. Why, then, is not every organism shaped like spaghetti? The reason lies in the wondrous ability of proteins to become molecular Erector sets for the building of some very intricate three-dimensional structures.

The ability of proteins to provide shape is twofold. The first is catalytic. If an organism has a certain enzyme to make a mucoid polysugar or the enzyme system to make hair, the shape and texture of its surface will be altered thereby. The second is inherent in the structure of the proteins: Their component amino acids have a built-in propensity for folding and crosslinking, thus enabling them to shape interlocking three-dimensional constructions. There are examples of proteins folding into their preordained three-dimensional shape.

For example, the hormone insulin is composed of two strands of amino acids linked together in specific sites via a special atomic ability of one of its component amino acids, cystine. We might symbolize these two strands of insulin A and B thus:

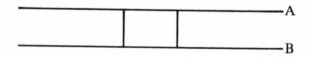

It is easy to see how the sequence of amino acids to form each of these strands is inscribed linearly in a filament of DNA. But, once formed, how do these two strands find each other amidst the millions of other protein strands being stamped out at the same time on the assembly line of the ribosomes? Random collisions? The formation of so important a substance as insulin, whose absence dooms the organism to instant coma and certain death, could not be entrusted to chance. The following mechanism was evolved: The structure of insulin is linearly inscribed in DNA; but the inscription is about 30 amino acids longer than the actual structure of insulin calls for. The two functional parts of insulin are separated by a sequence of innocuous amino acids as in the accompanying diagram.

A  B

This insert can now fold and bring the two precious parts in sufficient proximity to ensure the production of functional insulin by the formation of the molecular bridges and by the elimination of the inert insert.

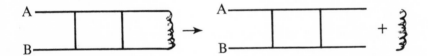

The excess energy required to achieve this is enormous: To inscribe the information for the 30 extra amino acids, 180 extra bases must be woven into the DNA gene (3 for each amino acid plus 3 to form the double strand of DNA). Ninety extra bases must be woven into the messenger RNA; 31 ligatures between the extra amino acids have to be formed. All of this—the synthesis of the bases, the synthesis of the amino acids, and their attachment into DNA, RNA, and protein—consumes energy. The squandering of this energy merely to ensure the meeting of strands A and B might appear wasteful; but any species of animal that tried to stint on this step is no longer with us. It disappeared for lack of insulin. Energy and molecular skill eliminate dependence on chance.

From Gurdon's experiment, then, we know that all the information of a species resides in the nucleus of every somatic cell. And we have just described the mechanism of the retrieval of that information by the synthesis of proteins from the stored information. The question we face today is: What is the mechanism of *selective* retrieval of that information so that only the proteins needed in a particular cell or at a particular stage in the development of an organism are synthesized? This mechanism, whatever it is, is the basis for the shaping of specific organisms via differentiation. Today we have some ideas, and some fragments of information about this process, which have yet to be woven into a meaningful pattern. It may well be that we are in a situation with respect to differentiation which resembles the condition of our knowledge of molecular biology prior to Avery's discovery. Most of the components of the system were known. DNA, ribosomes, and even transfer RNA were recognized as components of the cell. But it took us some 25 years to piece all of this information together into a unified whole. For all we know, some of the regulatory mechanisms of differentiation may also be known as components of the cell. But their precise function still eludes us.

We frequently hear criticism from some scientists who are appalled at the length of time that it took to decode the molecular biology of information storage and information recovery in a biological system. With the wisdom of hindsight, they suggest that they might have achieved all of this much sooner. We can offer them new challenges worthy of their mettle: Our knowledge of the expression of biological potential is but fragmentary; and our knowledge of the storage of *acquired* information, that is, learning by the brain and the recovery of that information, is totally obscure. These challenges have been with us for a long time, and I am certain will remain with us for a long time.

A living organism is an awesomely complex system; so much so that its very existence is highly improbable. To unravel its secrets requires ingenuity, dedication, and humility.

*Behold, how good and joyful a thing it is, brethren,
to dwell together in unity.*

BOOK OF COMMON PRAYER

Chapter two

The High Aspirations of the Lowly Slime Mold

Dictyostelium discoideum, a slime mold, is a unicellular, micro-scopic amebalike organism which was doomed eons ago to spend a solitary existence in dark, moist, detritus-laden soil such as the wet ground in a forest. (From now on let us call him Dd for short.) Life for such a creature is lonely and hazardous. Competition for food and survival is fierce. Voracious competitors, which gobble up available food and Dd himself, are myriad. Indeed, a hungry Dd may gobble up its kindred; for the sad truth is that our tiny friends may resort to cannibalism.

Escape from the predators or migration to greener pastures is rendered uncertain by the enormous distances involved. Dd may

33

be 5 microns in diameter; it would take perhaps 200 of them to reach across the dot on a letter "i." To move 12 inches, in terms of its body length, corresponds to a trip of 70 miles for *Homo sapiens*.

Evolution, that quixotic creator of diversity, pointed its finger at Dd and chose it for survival. It was to acquire the ability to cooperate with tens of thousands of its peers, to form large, motile colonies, to develop primitive eyes, so it could strive toward surface light, and finally to raise itself by its own bootstraps to huge heights—a quarter of an inch—and gain the advantage of scattering its spores from its great eminence. For it acquired this very advanced mechanism of reproduction, too.

Study of the life of Dd can be very instructive, for in this simple organism can be found forerunners of many of the systems which sculpture higher organisms, all of which start life, like Dd, as a single cell. We may start our observation of the organism when it is a unicellular protozoan. It looks like many other protozoa, it is microscopic in size, it has a nucleus; it has acquired a primitive apparatus for locomotion: flagella or pseudopodia. Since it lacks a photosynthetic apparatus, it must forage for its food. Dd is a predator.

If a bacterium happens to be in its vicinity, Dd stretches out part of its own tissue, forming primitive tentacles, which encircle the smaller, less motile, bacterium. The captive is not integrated into the cytoplasm of the protozoan, for the latter's cell membrane is never broken; Dd surrounds the captive and keeps it in a hollow compartment. Now some enzymes ooze out into the hollow: The bacterium is dismembered and the molecular debris is the reward of the successful hunt. Synthetic apparatus reshapes the DNA, RNA, and protein of the bacterium into the same molecular components characteristic of Dd. In due time (under ideal conditions "due time" is about 4 hours) the well-fed creature has grown enough and made enough cellular components for two cells. It goes through the sequence of intricate molecular maneuvers which I described in the earlier chapter and divides into two cells.

The offspring promptly start hunting on their own and the cycle is repeated. Occasionally two cells, via some molecular magic, are attracted to each other and in the privacy of their dark world embrace sexually and fuse into one large cell. The skin movies of the fusions, taken through the prying lens of the microscope, do not reveal whether this is an ecstatic experience but the biological consequences of such primitive sexual congress are known. Any mutational changes which had occurred in each of the mating cells or in their direct ancestors are pooled in the product of the fusion, or zygote, and some novel traits or abilities may become part of the hereditary endowment of the descendants of the zygote. Should these offer an advantage in the continuing or changing environment, the product of the sexual fusion may gain ascendancy over other cells, which are less fortunately endowed.

From our observation of Dd so far, we cannot detect any difference from other predatory protozoa. But somewhere within its genetic apparatus a tremendously important capability is inscribed. This ability expresses itself only when a drastic change occurs in the environment, when food runs out. What Dd does under these conditions is best observed not in its native habitat but under the controlled conditions of the laboratory.

We fill glass plates with agar solution liquefied by heat and incorporate into the agar the simple diet needed by bacteria, such as the bacillus found in our alimentary canal—the colon bacillus which has the redoubtable name *Escherichia coli,* or *E. coli.**
Escherichia coli, unlike Dd, can live on a very simple diet. It needs only salts and sugar, and from these simple ingredients it fashions the multitude of complex molecules, amino acids, fats, nitrogenous bases for its nucleic acids, and even all the vitamins it needs. The synthetic apparatus in this microscopic cell excels by far all of man's chemical factories.

When the agar cools, we spread upon it about a hundred million

* Mr. Escherich was a German bacteriologist whose monument is the name given to these organisms which are teeming in our colon. Scientists are often given unusual monuments. A famous, widely used strain of *E. coli* is one designated ML. The ML stands for Monsieur Lwoff, the great scientist of the Pasteur Institute who isolated ML from himself. We shall encounter Monsieur Lwoff again later, for his laboratory is one of the fountainheads of molecular biology.

living *E. coli* and perhaps a hundred Dd. The *E. coli* grow and multiply very rapidly, doubling every 60 minutes or so. Within 24 hours there is a visible whitish lawn of *E. coli* covering the whole plate. For the Dd this is a lawn of heaven. They feast on the teeming *E. coli* but, since the protozoan divides only once in 4 hours, their presence is not seen for another 24 hours or so. At that time about a hundred holes in the lawn become visible to the naked eye. If we scoop up a bit of moisture from the center of one of these holes and place it on a glass slide, a microscope will reveal to us a scene reminiscent of rush hour in New York: Hundreds of Dd rush about hunting for more *E. coli*.

The origin of the hundred holes in the lawn of *E. coli* now becomes clear. The pioneer Dd on the Petri plate began to feast in solitude on the *E. coli* amidst whom they found themselves. Each of them gorged and grew and divided in about 4 hours. The daughter cells promptly started foraging and repeated the process, and 4 hours later there were 4 grandchildren grazing on the *E. coli*. The process continued, and after 48 hours there were about 4000 descendants of each pioneer close to the site of the original landing. They moved outward in voracious quest for *E. coli*, producing a hole in the otherwise continuous lawn of *E. coli*.

If, instead of only 100 Dd, we placed on the Petri plate a thousand or more of them, then in 48 hours the picture would be entirely different. Instead of a hundred holes where the *E. coli* have been eaten away, there may be so many that the holes become confluent. Under these conditions, the plate will look not unlike a wet, muddy football field after a hotly contested game. What will happen subsequently can be better seen not on a Petri plate but under finely controlled conditions which have been developed recently. On a Petri plate there will be some areas where the *E. coli* lasted longer than in other areas, so that the nutritional condition of Dd will be different in different areas and the events we are about to witness might be staggered.

To synchronize the performance we do the following: We use an artificial filter with porosity such that the Dd will not go through, but the smaller *E. coli* will, and we place on the filter a large number of Dd, say about 10^7 of them. If we apply gentle suction, the *E. coli* will pass through the filter and the larger Dd will remain on top. But, just to make sure that all the *E. coli* are eliminated or that they cannot grow, we place the filter on a pad soaked in a drug, such as streptomycin, which is lethal to *E. coli* but innocuous to Dd.

We now have a fairly homogenous population of Dd which has had its last meal at about the same time. If we place these filters in a low-temperature incubator in the dark and observe them at hourly intervals, a fascinating scenario will unfold before our eyes. After about 2 hours the uniformly distributed Dd will give rise to little mounds, and these mounds can be visible to the naked eye. The microscope reveals how these mounds were formed: Thousands of individual Dd headed for a center focus, forming little mobile streams like streams of ants heading for their ant hill. In about 6 hours these little smooth mounds begin to have little ridges on top of them. By 10 hours the ridges coalesce into one, and a little turret starts rising. At hourly intervals there are changes in the shape of this rising projection. After 20 hours we can begin to distinguish a little stalk reaching upward. In 24 hours this stalk may have risen to its maximum height of 5 millimeters. On top of the stalk there is a sphere the contents of which we shall describe in a little while. These various shapes are seen in Figure 2.1, and Figure 2.2 is a copse of fully differentiated Dd.

How did these structures arise? Examination, not with the naked eye but with a microscope, will reveal a highly organized, complex structure. These mounds, which are visible to the naked eyes, are composed of perhaps as many as 100,000 amebae which migrated together once the *E. coli* were exhausted and then underwent a variety of organizational steps to produce the final terminal structure which we see in 24 hours. It is interesting to note that, though these cells have submerged their individuality to form these organized structures, enough of their individuality remains that, if we disrupt them up to a certain time by gentle grinding with

J. T. BONNER

Figure 2.1. Stages in Differentiation of the Slime Mold

fluid and return the de-colonized cells to a Petri plate laden with
E. coli, they abandon their newly acquired life style, they return
to rugged individualism and graze upon the *E. coli* in isolation.
It is quite obvious that it is hunger which foists on them the *e
pluribus unum* pattern of life.

If we pick off the sphere on the top of the stalk by touching
it to a glass rod, and add it to a little bit of water, we can spread

J. T. BONNER
Figure 2.2. A Copse of Fully Differentiated Slime Mold

this suspension on a new Petri plate covered with *E. coli*. We can follow with a microscope the fate of the small particles which are present in that sphere. The particle is about 6 microns long and about 3 microns in diameter. Unlike the original ameba, it is not motile. A few hours after these particles are placed on the *E. coli* bed there is a split down the side of the pill-like structures and an ameba bursts forth. The solitary creature starts the cycle all over again. If the encapsulated progenitor of an ameba is not placed in the proximity of food, it remains dormant indefinitely. It is a spore. Spores have been studied more intensively in bacteria, among some of which they are available in great abundance.

The spore is a special product of differentiation to ensure survival in a hostile environment. In order to form it, a bacterial cell undergoes a series of complex transformations. Several layers of membranes are formed, a huge concentration of calcium ion is hoarded; to neutralize the electrical charges on the calcium ion, a special oppositely charged organic molecule is fashioned and stored away. The spore is now ready to withstand heat, drought, and starvation. There are well-documented demonstrations of the viability of bacterial spores after 50 years of dormancy.

The resistance of spores to killing by heat has caused the death of countless men feasting on presumably well-prepared food. Berton Roueché, that matchless chronicler of medical detection, described a case of lethal poisoning by botulin in an Italian family during a Thanksgiving dinner. They ate marinated mushrooms. The recipe for the dish is as follows: Wash the mushrooms in cold water, boil in white wine for 30 minutes. Place in a mason jar, season, cover with olive oil, seal. Because white wine contains 12% alcohol, its boiling point is well below 100° C. A half-hour of such exposure is just about enough to tickle the spores of *Clostridium botulinum* to emergence. It takes 6 hours of boiling at 100° C to destroy the spores of *Clostridium botulinum*.

Dd has means other than spores to facilitate survival. It can move. It can move to greener pastures. Its method of locomotion was first observed by John T. Bonner, a microbiologist at Princeton University, who has contributed much to our knowledge of the life and times of the slime mold. He performed the following experiment. Instead of spreading 10^6 or 10^7 Dd on a large surface he put them in one streak at one end of little boxes. Some boxes were covered so it would be dark underneath. A source of light was introduced at one end of the other boxes; this is done very simply by covering the little box with aluminum foil and punching a pinhole at one end of the foil. (This is a blessed area of biology where one can still do experiments with relatively simple and inexpensive apparatus.)

If after several hours the cover was removed from one of the boxes which permitted no light, aggregates of the slime mold which assumed shapes of slugs were revealed all over the field. The boxes in which a pinhole permitted light to penetrate revealed a fascinating pattern. The slugs all moved in the direction of the pinhole. Therefore they have the capacity to respond to light. After the opaque cover was removed, the slugs stopped moving and began promptly to raise stalks and produce spores. We may presume that, for these slugs which were migrating randomly, presumably toward the surface of the substratum where they would normally be living, the light acted as a stimulus for halting their random march and for the production of spores. Dd is thus revealed as a very complex organism. It can aggregate. The aggregates can form specialized tissues such as stalk, spore capsules, and spores. The aggregates can move, and they can "see."

The slime mold can serve as an extraordinarily useful model for studying differentiation. Every other complex organism starts as a single fertilized cell. This cell divides and proliferates, and some of the mass of cells in the embryo give rise to certain organs. As we shall see in future chapters, this makes their study very difficult. With Dd, however, we have an entirely different situation. A hundred thousand cells, apparently completely alike, and with equal capabilities, decide, if we may be anthropomorphic, to move together and to undergo certain profound alterations for some common goals. What are the biological mechanisms which achieve this remarkable transformation, which might be considered a form of differentiation?

The first question we may ask is, What is the signal which causes the ameba to concentrate in one site? After much tortuous research, the nature of the signal was revealed. It is a molecular signal which, interestingly enough, is an ancillary agent of hormones, the regulatory agents of higher organisms. Early workers studying this slime mold observed that, if some fluid is taken from the focus to which the protozoa have already streamed and this fluid is placed among protozoa which are not yet aggregating, they promptly start movement toward the transplanted fluid. The agent in the transplanted fluid which may be responsible for this movement was given the name acrasin from the generic Latin name for these organisms, *Acrasalia*.

Biologists are quick to give names to hypothetical substances. Some of these disappear into limbo, but the reality of others can be established, provided certain circumstances exist. In the first place there must be an assay for the substance. Then we can follow and possibly concentrate the material which is the putative cause of the biological phenomenon. The hypothetical substance must be stable, otherwise it remains elusive.

In the identification of acrasin there was a fortuitous circumstance; the material had been known and some of its ancillary functions with hormones had been well defined about 8 or 10 years before. The biochemist, Sutherland, about whom more in Chapter 5, had observed that for the activity of some hormones in animals a small activator molecule is needed. He identified this substance as a very unusual molecule which we call cyclic 3', 5'-adenosine monophosphate. As it turned out, this substance, whose pet name in the laboratory is cyclic AMP, is the effector for many different hormones. That acrasin is simply cyclic AMP also was clearly demonstrated in the following way. A large number of Dd were harvested and, after they began their period of starvation, were placed on membranes which permitted the flow of small molecules through them. The slime mold began to aggregate on the membrane, and the fluid which seeped through was collected beneath the membrane. This fluid, when added to nonaggregating Dd, promptly caused their aggregation. The material was concentrated, isolated, and a variety of molecular fingerprints proved it to be cyclic AMP.

The next obvious question is: Why does cyclic AMP not accumulate in amebae which are feeding? The curious answer is that while they are feeding the amebae also excrete an enzyme which destroys cyclic AMP. Thus this is a very primitive regulatory mechanism. During periods of plenty, the amebae can synthesize an enzyme; for this task, energy and a large number of amino acids are available. At the same time cyclic AMP is also being produced and excreted into the surrounding fluid and is promptly destroyed by the enzyme. Now comes the period of starvation; without food, Dd lacks the means, both material and energetic, to continue the production of the large protein—enzyme. Starvation then permits the accumulation of cyclic AMP in the medium, and this substance then becomes the chemical signal indicating that it is time to aggregate.

This seems to be a lugubriously complicated system for notifying Dd that he is starving and he should seek whatever advantages aggregation may bring. But, then, we must learn never to question the intricacy of the mechanisms by which evolution achieves its purpose. Evolution is a pragmatic empiricist with a strong addiction for gambling. It throws its molecular dice a million times and tries out the permutations that are produced. The useful permutations are retained and perpetuated. The others are cast aside.

Evolution has endowed Dd with a variety of other enzymes for the efficient completion of its life style. It is obvious that new enzymes are needed. For the ease of its migration, Dd excretes a slimy material, hence the name slime mold.* Such a material is made of a polysugar for the synthesis of which new enzymes are needed. The stems on which Dd grows its spore capsule are made of semirigid cellulose. For the synthesis of these, too, brand new enzymes are needed.

It is fruitless to speculate on exactly how the information for these enzymes has become part of the genetic endowment of Dd; but we can fruitfully study the synthesis of these enzymes in current, modern-day organisms, for molecular biology has given us the tools for such an examination. If we measure the total RNA in, say,

---

* It is certain that the reader has been annoyed sometime by the propensity of the slime molds to produce a slippery slime. Slime molds can thrive at low temperatures. Refrigerators containing meats are often infested with them, and their growth in meats stored for long periods is unmistakable.

10^8 cells at the beginning of aggregation, that is at 0 hours, and 20 hours later, we will find that the total RNA of the whole mass drops to less than one-half of that present originally. The total protein also drops to just about half of the protein content of the individual unaggregated amebas. It is obvious that during this period, which is a period of starvation for the organisms, these two large cellular components are being used for some purpose. However, even though the total amount of protein and the total amount of RNA are rapidly decreasing, new protein and new RNA are being synthesized during the period.

We can prove this with the aid of radioactive tracers. If during aggregation we provide for the starving organisms amino acids which are labeled with the isotope of carbon—carbon^{14}—which is radioactive, we can find that the radioactivity, and therefore the amino acids, are being incorporated into proteins. The same thing is true with respect to RNA. Even though the total amount is diminishing, we can show with our radioactive tools that there is new synthesis.

Why all this destruction of old and synthesis of new protein? We can actually show that brand new enzyme activities which are absent from the amebas appear at definite intervals during the period of differentiation. For example, the enzymes that are needed to make the polysugars for the excretion of the slimy material can be shown to arrive about 7 or 8 hours after the beginning of aggregation.

Where is the information for the synthesis of these new enzymes? We can, of course, assume *a priori* that the information is present somewhere in the DNA. But we can demonstrate this experimentally. A drug, actinomycin D, is known to inhibit the transcription of DNA into messenger RNA. This drug binds to the DNA in such a way as to prevent the enzymes which transcribe DNA into RNA from coming into contact with DNA. If actinomycin D is added to the amebae during the first 3 hours of aggregation, the amebae do not go through the usual structural alterations. No slug, no spores are formed. If the drug is added 3 hours after the beginning of the aggregation, the slug, the stalk, and spores are synthesized. Therefore all of the information for the production of these highly complex structures must be present in the DNA, and the information must be released during the first 3 hours of aggregation.

It might be objected that the drug actinomycin D does not penetrate well into the complex *after* the aggregation. This objection can be met and countered the following way. After 10 or 12 hours the organisms can be disrupted very gently down to the single-cell stage and returned to the filter and incubated in the presence of actinomycin D again. Essentially what we have now is unicellular organisms, which have experienced aggregation previously and now are in the presence of the drug as individual cells. They will aggregate again and go through all of the stages of differentiation, including the production of spores. Therefore, having eliminated the objection of the nondiffusion of the drug, we may conclude that the behavior of the organisms during the period beyond the administration of the drug has been programmed into them prior to the administration of the drug.

There have also been intensive studies on the effects on the development of the slime mold by drugs that prevent protein synthesis. If we add such a drug after 6 or 8 hours of incubation, all subsequent development is inhibited. In other words, some proteins must be made beyond this point for the development of unique structures. However, if the drug is removed, the amebae go through their normal aggregating development. This again implies that the information for the making of the new proteins is present, preprogrammed and, if the protein-inhibitor drug is removed and protein synthesis is permitted to proceed, normal development follows.

With these studies we have established the slime mold as a valid example of a differentiating organism. Its cells, after a certain stage, assume specialized roles, such as stalk cells, making only some molecular structures, such as those which confer rigidity, and serve the whole organism. Moreover, once a cell is committed to a specialized structure and function, the process is irreversible: A stalk cell cannot start life again as a free-living ameba.

The evidence is thus clear that for the shaping of structures there must be new messenger RNA synthesis and new protein synthesis as well. How did these new capabilities of producing new proteins arise? Molecular biology has no satisfactory answer. We understand how mutations occur in preexisting proteins. For example, there is an amino acid, alanine, in position 38 in the insulin of cattle which differs from the amino acid in the same position in the insulin of the pig, in that insulin, amino acid 38 is threonine. The codes for these amino acids in DNA are CGA and TGA, respectively. It is easy to see that an exchange of the initial base in these two triplets could account for the mutational change in the insulin of the two different species of mammals. Fortunately, that mutation was not deleterious to either species, so we can have porkchops *and* steak!

Very few such amino acid exchanges are known in the insulins of different species of organisms. The reason for this presumably is that mutations in amino acids which are critical to the functioning of insulin are lethal and could not be perpetuated.

But the Devil's Advocate or, more appropriately, God's Advocate if we may indulge in such belittling blasphemy, may confront the molecular biologist with the problem: Now explain the origin not of a mutation in a preexisting gene, but the origin of a brand new segment of DNA which specifies the structure of a never-before-existent protein. There are two different sets of contrived answers to this question.

The biochemists who are preoccupied with prebiotic molecular evolution, many of whom are committed to an absolutely reductionist philosophy, have a glib answer. Let us assume that some primordial blob of jelly which was acquiring amino acids from the vast primordial sea ran out of one of the needed amino acids. If such a creature had acquired the enzyme to make that amino acid with the aid of sunlight from some simpler molecule, it would have had an enormous advantage for survival over its jelly peers without such enzymatic endowment.

This is simple to state, but its molecular probability is vanishingly low. Let us assume such an enzyme to have 100 amino acids. For encoding a protein of such length, 300 nucleotides would have to be aligned in exact order in DNA. Each triplet would have to be in absolute order and, in turn, the 100 triplets would have to be in absolute order. The random probability of this event can be calculated as follows: The probability of any one triplet having the right permutation is 1 in 64. The recurrence of the appropriate triplets a hundred times has a probability of 1 in 10^{100}. A chance of $1/64 \times 10^{-100}$ is no chance. Nor is it helpful to assume that 100 amino acids were willy-nilly lined up and fused into the felicitous sequence and the nucleic acid sequence was shaped on *it*. The probability of *that* occurrence is 10^{-100}. But molecular evolution had a long, long time for the random throws of dice in the form of amino acids.

Another equally improbable hypothesis states that the primordial organisms had *all* the synthetic abilities and through eons of evolution lost some of them. This, of course, amounts to Creation by some omnipotent force. The probability of each of the schemes is so small they require a leap to faith. The leap in one direction is as justified as in the other.

The committed molecular biologist has enough on his hand decoding contemporary mechanisms without probing primeval systems which left no trace. The origin of a new sequence of DNA and of the ensuing protein remains a presently impenetrable mystery.

Always the procreant urge of the world.
WALT WHITMAN

Chapter three

Sex

Sex, which is merely fun for the individual, is indispensable for the species. Without the pooling of genes of two individuals of opposite sexual polarity, evolution would have been so slow we might still be wriggling worms in a vast sea of algae.

A distinguished geneticist, Werner Braun, looks upon sex—most unromantically—as a defense mechanism. It is a defense mechanism of the species against changing environment, enabling some members of the species to adapt and survive.

How very important sexual union and the consequent fusing of genes is to the species is best documented by the astonishingly complex and ingenious methods which have been evolved to ensure meeting and mating of the appropriate individuals. Probably the most sophisticated mechanism to this end is that developed by insects.

49

Consider the problem of a female winged insect. She has an urgency to lay her accumulated eggs, which may number many hundreds. Her flight is impeded by her heavy burden; males of the species, buffeted by winds or impelled by their male fancy, may be miles away. To rescue the female from her predicament, Evolution endowed her with a signaling device of extraordinary sensitivity. Female insects can exude odoriferous chemicals specific for that species. These molecules are wafted by the winds, male insects miles away pick up the molecular signal via an olfactory apparatus in their antennae, and lo! as by the voice of the turtle, they are beckoned to pleasant duties. The males will fly unerringly toward the origin of the scent. They will struggle against the wind and, if they should lose the trail, they will fly crosswind back and forth until they pick up the scent again and can follow the nuptial trail to its promising source.

The extraordinary potency of these substances can best be judged by some studies which were initiated not by voyeur interest but by practical consideration of insect pest control. Male pest insects can be destroyed by baiting traps for them by artificially made olfactory ambrosia. This sneaky business is carried on extensively as follows. Some 500,000 virgin female insect pests are deprived of their abdominal tips, which house the attractant, or pheromone. The term pheromone has been coined for such substances which, unlike a hormone, are excreted outside the organism's body to achieve some purpose on another organism. The tissue of the insect is macerated, extracted with a variety of solvents, and the pheromone is concentrated. The guide in the processes of concentration is the response of the male insect to the product, at great dilutions. Finally the product is concentrated enough to be purified, analyzed for its structure, and synthesized in the laboratory by the organic chemist.

With the inexpensive synthetic material in hand, control of the insect pest becomes a simple task. Traps are baited with it and millions of males are cajoled by the siren smell to their doom. The efficacy of the lure would be unbelievable if it were not documented by well-controlled field studies. If 10^{-9} gram of the specific pheromone of the gypsy moth is placed in a trap, the first week it catches 70, six weeks later 20, and 12 weeks later 12 of the hapless males in quest for love. It is difficult to visualize 10^{-9} gram (a dime weighs about 2 grams so we must visualize a mass equal to one billionth part of the dime), and yet this trace amount of sex attractant keeps giving off enough molecules for 12 weeks to retain its potency. This kind of research, which appeared rather esoteric decades ago, can yield a rich bounty. As we slowly realize that covering our earth with insecticides is devastating our total ecology, we must turn to more specific techniques for insect control.

One scheme which has been tried on a small scale and found promising is devilish. The male-attracting pheromone of an insect pest is scattered to blanket an area completely. Since the amounts needed are small and the chemicals cheap, this is a feasible approach. The males go mad with excitement and fly off in all directions and, since their quest leads nowhere, they continue their frenetic search and drop dead of exhaustion. The miniscule amounts of insect pheromones released into the atmosphere harms no one except the male insects and the unfertilized eggs.

Some primates such as the rhesus monkey are known to have sex pheromones. Whether human beings produce them individually is a moot question; but their collective use supports the huge perfume industry.

Aquatic animals also use pheromones and other signaling devices to make their availability known. The female lobster entices its male with an as yet unidentified pheromone released into the water. Deep sea fish that swim in eternal darkness signalize their species individuality by characteristic patterns of illumination produced by light-generating enzymes. Some fish have a special blushing organ which suffuses them with a pink color during receptive courtship (Figure 3.1).

The methods of bringing sperm and egg into close proximity are as varied as those which entice the sexual mates to closeness.

To ensure probability of sperm meeting egg, some organisms such as fish produce prodigious numbers of both; the caviar industry is a by-product of this prodigality. In many instances, provision has been made by evolution to introduce the sperm into the body of the female, or *vice versa*, as in the sea horse, that organism which is the triumph of the piscine female lib movement since the male carries the fertilized egg!

Such docile acquiescence on the part of the male is rare: the mating pattern of the salamander is more representative. Once a female catches his fancy, the male salamander courts her by clasping her and rubbing her nose with his chin. If he receives some signal of acquiescence, he dismounts and deposits in front of her a sac full of his sperm. The female then picks up this bundle and deposits it into the appropriate aperture in her body. The sperm sac breaks and fertilization ensues.

Where no apparatus for achieving such proximity exists, some aquatic organisms have agents to induce the male to void its sperm soon after the female voids its eggs. Those of us who have worked on sea urchin embryogenesis—our number is legion—have cursed in many languages this particular pheromone. You have an idea about some facet of embryonic development and you are fiercely eager to test it. You order a couple of dozen sea urchins from a supplier in California or Florida. The airplane with the shipment arrives invariably at midnight, but you yourself meet it and take your precious cargo to your laboratory. You rouse the guards to let you in, you carefully disentangle the sea urchins from the seaweed (in your eagerness you are punctured a couple of times) and gently place the animals in tanks filled with aerated artificial seawater maintained at a temperature salubrious to sea urchins. You return home where your wife mumbles in her sleep that next time she will marry a fireman because then at least she will know what he is doing at night.

Next day on the way to the laboratory you visualize the beautiful experiment. You will inject some potassium chloride into each sea urchin, place it on top of a beaker filled with saline solution, and watch the sperm or eggs ooze out, enabling you to harvest

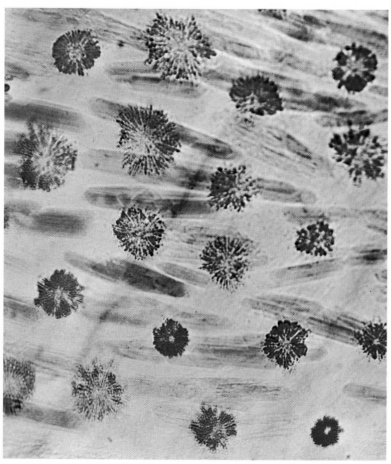

3.1 The Blushing Organ of the Fish

sperm and egg separately and perform synchronous fertilization and subsequently make a brilliant finding never before observed by the eyes or mind of man. Finally, when you approach the tank and you see not clear seawater and sea urchins awaiting you but a white cloudy mess, your nurtured dreams are buried to the accompaniment of a dirge of words which I shall not reproduce here. What happened was that one of the female sea urchins, either because of the novel environment or because her time had come, let go of her hoard of unfertilized eggs, thus stimulating every other female to do the same and, in turn, every male got into the act; and that is the end of that experiment and that dream.

The mass spawning response to such individual stimulus is put to good use in the oyster industry: By throwing female oyster gonads into oyster beds, the oyster farmer induces mass spawning and finds that he harvests better yields of oysters.

Let us now turn to the origin of the sex cells in which all hereditary information is placed and whose fusion is the goal of highest priority of every living organism.

In ordinary, or somatic, cells there are pairs of chromosomes, the members of each pair being quite similar in appearance and presumably in function. In the human, 46 chromosomes are made up of 23 pairs: Such a cell, containing the full complement of chromosomes, is called diploid. As we pointed out earlier, if sexual union fused two diploid cells, the chromosome number in each generation would double, requiring in not too many generations a cell the size of a football stadium to house the chromosomes alone. Such a calamity could be avoided by one of two ways. Chromosomes could be discarded at random. (This is essentially what happens when two cells with different chromosome numbers are fused artificially, as we shall see in the next chapter.) No living organism could survive such a chaotic system: If chromosomes bearing pivotal information were jettisoned, that offspring would be doomed to death. Another method of maintaining the correct chromosome number is to reduce the number of chromosomes to one half in every sex cell.

The process of special cell duplication by which the number

of chromosomes is reduced by one half is called meiosis, and specialized organs were evolved to achieve it. Spermatozoa are produced continuously in tubules within the testes by a very complex sequence of cellular events. In ordinary mitosis, during prophase, there is essentially a doubling of the chromosomes, ensuring that each daughter cell will be endowed with the full complement of chromosomes.

During meiosis, on the other hand, there is not one but two cell divisions but only *one* chromosomal division. Thus in meiosis during the prophase there is a normal division, but now in quick succession two more cell divisions occur, so that the chromosomes are distributed into four cells, each of which receives, therefore, only one-half of the chromosomes of the species. After this a very elaborate construction sequence ensues. The future sperm cells are actually adopted by feeder cells which apparently nourish the parasitic sperm, developing it for its mission.

When early biologists looked at the sperm cell through a light microscope they saw an apparently simple tadpolelike structure. But what a complex little missile was revealed once we focused on the sperm the searching beam of an electron microscope (Figures 3.2, 3.3).

A typical sperm has three main parts: the head, the middle structure, and the tail. Each has highly specific functions and appropriate molecular equipment for fulfilling its task. Thus the head has several sheaths to contain its three main components. The most anterior snoutlike structure, called the acrosome, contains a special enzyme called lysozyme (or dissolving enzyme) to whose molecular attack, as we shall see later, a part of the wall of the egg yields, and the sperm is thus permitted to penetrate into the center of the egg. The largest part of the head is the nucleus, which contains the genetic material that is to be introduced into the egg. The chromosomes are very tightly packed. This is achieved by extruding water from the head, the water being bailed out by an exquisitely selective machinery to reduce dead weight and thereby increase the potential

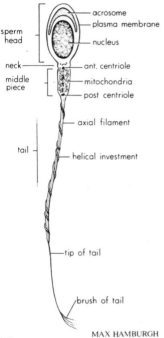

acrosome
plasma membrane
sperm head
nucleus
neck
ant. centriole
middle piece
mitochondria
post centriole
axial filament
tail
helical investment
tip of tail
brush of tail

Figure 3.2. Structure of a Sperm MAX HAMBURGH

range of the missile. Even all the RNA which is abundant in all other nuclei, especially in nucleoli, is jettisoned.

During this preparation for the journey the shape of the head changes; with prescient knowledge of hydrodynamics it flattens in man and bull, or may become spirally twisted like a corkscrew in birds, or it may become a pointed scimitar in rodents. These are the shapes naval engineers might design for more efficient propulsion in water.

Still another important structural feature of the sperm is a small granule at the base of the head, the centriole, which is carried along with immense foresight for it will be essential for the initiation of cell division in the fertilized egg.

The middle piece of the sperm is essentially the motor of the torpedo. It is packed with mitochondria, which are the powerhouse

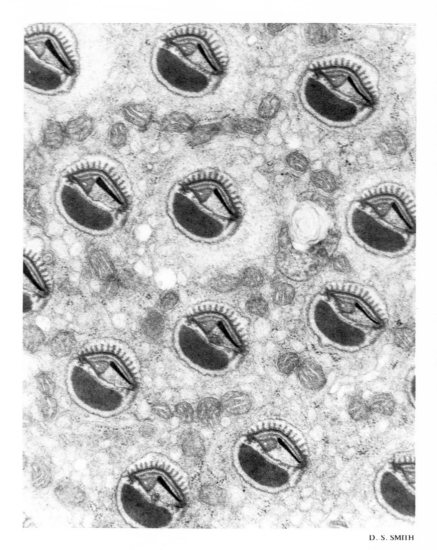

D. S. SMITH

Figure 3.3. Insect Sperms in Cross Section

of life, converting the energy stored in glucose into biologically useful energy contained in a special molecule, ATP, the incredibly versatile battery fueling every energy need of biological systems —heat, light or, as in this case, motion.

The tail is, of course, the propeller. It is structured for that specific task along essentially the same lines as similar organelles of locomotion, such as the flagella of protozoa. This is one more example of the frugality of Nature in its inventions. The flagellum may have been evolved in some primordial asexual motile creature, and its essential features were incorporated to give motility to sperm. There is a pair of fibers in the middle of the tail and another ring of fibers surrounds the two. These fibers contain a special kind of protein, also found in muscles of animals, which can contract and expand. The mechanism of such rhythmic motion was revealed by the brilliant work of Albert Szent-Györgyi, a Hungarian biochemist, who gently removed the functional proteins from the structural fibers of muscle and was able to show that these proteins contract *in vitro* upon the addition of ATP and expand once the energy of ATP is spent.

This demonstration was a milestone in the progress of biological knowledge, for motion is a hallmark of the living and its mechanism was revealed to be subject to the laws of molecular mechanics: one more triumph of the reductionist approach to the study of life. Very soon after Szent-Györgyi's demonstration an investigator extracted similar contractile proteins from sperm and, still more extraordinary, an investigator who had the skill and infinite patience showed that even the bacteriophage has the same self-propelling mechanism.

The contractile protein of sperm is actually a much more rewarding object of study than the original muscle protein studied by Szent-Györgyi. Those proteins, when extracted and exposed to ATP, simply contract, and stay that way until the ATP runs out or they are placed in a fresh medium devoid of this ubiquitous source of energy. On the other hand, the contractile proteins from the bull sperm, for example, when placed in contact with physiological amounts of ATP, begin a rhythmic wave formation which may last for a couple of hours. Both the amplitude and the rhythm of beat are quite comparable to that of live sperm. However, *in vitro* there is no coordination of the wave motion: The muscles are aimless. The orientation mechanism of the intact sperm is still unknown.

The intact sperm with its tail vibrating in a whiplike motion heads for its target like a well-aimed torpedo. This sentence is but a figure of speech, for there is no evidence for long-range signals between egg and sperm as there are between the male and female whole organisms to lead them together. The encounter is probably achieved only by the production of vast numbers of sperms and by the use of an appropriate organ to launch them in the vicinity of the eggs.

There was some disagreement among biologists whether the sperm tail propels itself with a corkscrewlike motion or by merely vibrating in one plane. Whatever the swimming stroke, circular or lateral, the sperm heads toward its target.

For shaping the egg, even more intricate biological maneuverings are required than for the production of the sperm. For the functions of the egg exceed by far the single-track role of the sperm. The egg must not only contain one half of the chromosomes, it must also have hoarded in it enough fats, proteins, sugars, and minerals to nourish the embryo until hatching in nonmammals, or until communication with the mother's bloodstream can be established in the mammal. Thus, for example, the immature frog egg during a period of 3 years of development increases in size about 30,000-fold. In the mammal, on the other hand, all the eggs which will mature in the female are present in primitive form at birth and will increase only about 40- to 50-fold in mass during development, which may take a few weeks, or as long as 40 years, as in the human.

For the shaping of the egg, evolution might have developed two different processes. After meoisis, as in the development of the sperm, the four eggs with the haploid number of chromosomes might have been permitted to accumulate the required foodstuffs and thus four eggs would be ready to perpetuate the species.

However, Evolution decided on a different pathway. Food, such as yolk, and the proteins of egg white are accumulated prior to meiosis. This is a very complex process involving organs far from the site of egg formation. For example, in vertebrates the yolk components are synthesized in the liver of the female and carried by the blood to the site of the developing egg, where they are deposited as yolk granules.

Hormones have a stellar role in this process, which will be discussed in a later chapter, but as evidence for the participation of hormones it might be cited here that *roosters*, too, can be induced to produce egg yolk proteins in their liver by the injection of huge doses of the appropriate female hormone! This manipulation, which sounds like a dirty trick to play on the hapless rooster, is very revealing of regulatory mechanisms.

After the egg grows to its requisite size, complete with nucleus, yolk, and other foods, it undergoes two reduction divisions as does the developing sperm. But during these divisions the yolk and other food components are partitioned very unevenly, the bulk being sequestered in only one egg. As a result, the end product consists of one large haploid egg endowed with food for its future tasks and three puny cells, also haploid, which, however, bereft of the hoard of food, have no place to go, so they die.

The reason for this apparent waste can only be a logistic one. Before the reduction division the female would have to carry a prototype egg which is four times the size of the final egg. Since in some animals such as the shark, or the ostrich, the egg can be very large, this could be a huge lump, perhaps 8 pounds in the ostrich, which would be a logistic and esthetic handicap. This, of course, does not apply to the mammal, whose eggs are very small. But this is really a small waste; there was no compelling reason for remedying it by the invention of a new process of development of the egg.

In addition to the nucleus, yolk, and white, eggs are endowed with a system of membranes—in the birds there are no fewer than five distinct layers. Since the reader has seen some of these many times at breakfast, they need not be listed; however, something that is not so readily visible should be mentioned. The egg shell is not only a protective device against mechanical damage, but it serves also as a barrier against unwanted external gases and it is a filtering device to permit escape of internal gases. A hen's egg contains about 7000 miniscule pores through which oxygen is inhaled and carbon dioxide, but not water, is expelled. Otherwise, the egg is a closed system, a universe unto itself, in which a miraculous transformation of its contents, from blobs of yolk and white to a living creature of flesh, blood, and bones occurs.

How efficient the material transformation is can be seen from the following figures. The total amount of nitrogen in a frog egg is 2×10^{-4} gram, the total nitrogen in a newly hatched tadpole is also 2×10^{-4} gram. The total sugars, however, are reduced to one-half; these were consumed for energy and lost as carbon dioxide. The weight of the newly hatched organism may be greater than that of the egg owing to absorption of water, but such increments in weight are modest except in viviparous organisms in which the fertilized egg establishes a direct connection with the maternal body via a lifeline, the placenta, a marvelous organ which was invented by Evolution several times: In addition to placental mammals some very primitive contemporary organisms such as tunicates and some sharks have it too.

Let us return to the sperm about to be launched for its mission. Evolution could not attain the ideal and most economical mode of fertilization, which, of course, would be a 1:1 ratio between egg and sperm. Failing the achievement of that kind of material efficiency, a brute-force approach was developed.

To ensure fertilization of a single egg vast numbers of sperm are shaped and launched. The human male, for example, produces about 3 cc of ejaculate which contain about 180 million sperm. From studies of human fertility we know that an effective sperm count is about 50 million per cubic centimeter. If the cell count falls to 10 million per cubic centimeter, the probability of fertilization is poor. There is a very large variation among individual males, and it is even larger in the sperm productivity of different species of mammals. The bat's ejaculation may have a volume of 0.05 cc and the boar's may reach ten thousand times that volume, 500 cc. In the latter, the sperm count may be 10^8 sperm per cubic centimeter or a total of 5×10^{10}, which is about twenty times the total human population of the earth. How the boar has material and energy left for anything else is difficult to see.

Once the sperm are launched, they begin wriggling their way toward their target, which in the case of the human is a speck of jelly 0.14 millimeter in diameter. The pace is furious, about 0.1 mm per second—the need for urgency is real, the lifespan of a human egg is about 24 hours; the lifespan of the sperm is the same; the distance to be covered is great,* parts of the traveled tract are hostile; sudden changes in acidity may be encountered. The success of the sperm may be measured in an experimental animal. Of the tens of millions of sperm launched by a male rabbit only 50 to 250 may reach their goal, the upper end of the oviduct and the eggs, in 3 hours.

Although no long-range forces appear to exist to attract the sperm to its target, once the sperm is within the vicinity of the egg it encounters welcoming substances. It was the Canadian-American scientist F. R. Lillie who first showed that, in the presence of ripe eggs of the same species, spermatozoa become sticky and adhere to the egg and even to each other. Lillie called this molecular fly trap fertilizin. Extensive studies of fertilizin revealed its source to be the layer of jelly that surrounds the egg and is essentially fertilizin, which slowly dissolves, surrounding the egg with a halo of glutinous solution. Fertilizin is a protein combined with sugar molecules; such a protein is invariably sticky. Both the component

*The late George Gamow, that great popularizer of science, calculated the relative length of the trip that a human sperm must travel from the site where it lands to its goal, the egg, to be equal to a trip on foot across the width of the North American continent.

amino acids and sugars of fertilizin vary from species to species; thus the egg can be highly selective in the sperm it traps.

The surface layer of the sperm contains another protein to which the unfortunate name, antifertilizin, has been given, implying semantically that its role is to defy or destroy the fertilizin. Actually, the antifertilizin is part of the fertilizin machinery. These two proteins combine in a highly specific manner, as do enzymes with their specific substrates, or antibodies with specific antigens. Reactions between fertilizin and antifertilizin of different species are sometimes possible, provided the species are closely related to each other, but such bonds are much weaker than those between the homologous pair of egg and sperm. With such elaborate molecular machinery does Nature bar promiscuity, for an egg fertilized by a foreign sperm could only produce a monster should the fusion give rise to viable progeny.

There is most likely a second line of defense against the fusion of the male and female genetic material of different species. This is the molecular mechanism which stamps individuality on the DNA of different species. As stated in Chapter One, enzymes attach small molecular structures—methyl groups—to DNA in a pattern characteristic of each species. Certain other enzymes which can cleave DNA, thereby destroying its integrity, can recognize the novel pattern of methyl groups in a foreign DNA and promptly dismember it. The subtlety of this molecular screening is such that in one case appropriate enzymes can recognize the presence of 1 methyl group among 3000 bases of DNA.

As yet we have no experimental proof that this recognition is effective for the prohibition of fusion during fertilization of unrelated DNAs. But we will await the emergence of such evidence confidently, albeit patiently. Nearly 10 years elapsed between the suggestion and the confirmation that methyl groups serve in bacteria as protective signals from indigenous enzymes and signals for attack by foreign enzymes.

Even more subtle recognition mechanisms which permit only selected matings exist. Lord Rothschild, one of the foremost experts on the biology of fertilization, cites the example of certain forsythia that have two kinds of flowers which grow on different bushes. In one the styles are short and stamens long; in the other the styles are long and stamens short. In Nature, fertilization occurs only when pollen from a flower with long stamens comes in contact with a stigma on a long style or when pollen from short stamens comes in contact with a stigma on a short style on another flower. These different flowers contain individual small molecular components which presumably serve as recognition signals by some unknown process for fertilization by the appropriate pollen. In turn, self-fertilization is excluded by an equally unknown molecular fence.*

As the sperm is trapped by molecular forces between itself and the egg, the shape of the sperm becomes altered. The acrosome collapses and essentially disintegrates, releasing the egg-dissolving enzymes. The egg responds promptly by bulging forward at the point of contact; this may be a mechanical response from molecular forces within the egg. If we assume that the egg's contents are under slight pressure, then a slight break of the enfolding membranes would produce a forward thrust. This projection forms on both sides of the sperm's head, producing a structure known as the fertilization cone, which gradually engulfs the sperm and then, by retracting, drags its captive inward.

This event is the signal for the beginning of almost frenzied molecular activity within the egg. Before fertilization, the egg, though not completely inert metabolically, is essentially quiescent; its oxygen consumption is low. It is synthesizing practically no

---

*Lest the reader gather the impression that Lord Rothschild is a trivial individual content to be preoccupied with such minutiae, we must relate his wartime occupation. He was one of those few supremely cool Englishmen who would descend into the crater of an unexploded bomb. The cause of the failure to detonate was unknown, the fuse might have malfunctioned; or the Nazis were not above purposely arming bombs with delayed fuses as an added bit of terror. The image of these Englishmen, among them Lord Rothschild, astride a bomb containing tons of high explosive, calmly telling via a walkie-talkie what screw or what pin they were manipulating is awesome and, it is hoped, will be applauded as long as there are admirers of the triumph of human spirit and courage over marauding brutishness.

proteins. The latter we know from applications of the exquisitely sensitive isotopic tracer methodology. Minute quantities of amino acids containing one or more radioactive atoms are added to the fluid which bathes unfertilized eggs.

As mentioned earlier, the sea urchin is the favorite experimental animal for such studies. The radioactive amino acids pass into the interior of the eggs; we know this because, if eggs are subsequently well washed to remove externally sticking amino acids, and the interior of the eggs is extracted, the internal amino acid "pool," as it is called, is radioactive. By appropriate chemical manipulations we can separate the proteins from the "pool" and find them to be but negligibly radioactive, indicating that there was minimal protein synthesis. A similar experiment after the admixture of sperm and eggs will yield proteins in the fertilized eggs which will make the counting machine blink like a telephone switchboard on Mother's Day. It is obvious that fertilization is the signal for the production of new proteins needed for the manifold activities which lie ahead.

But let us return to the sperm which is just being drawn into the egg by the fertilization cone. There is great variation from species to species in how much of the sperm is taken in. In some species the egg pulls in everything, head, center, and tail; in others the tail and even the center piece are left outside. But in all cases the head enters, providing additional evidence, if that were needed, that the critical information is encased in the head.

What happens next also varies, depending on the state of the reduction division in the egg. In some species this is completed only after fertilization. In the sea urchin, in which the reduction division is completed before fertilization, the sperm nucleus proceeds to join the egg nucleus immediately. To do this the sperm nucleus, which we might now call the male pronucleus, goes through a well-programmed choreography: It gyrates until its posterior end is thrust forward, it swells by imbibing water from the egg, and all the while it is moving through the egg cytoplasm toward its predestined rendezvous with the female pronucleus, which also begins to move toward its mate.

Fusion of the two in different species also varies in minor details; but essentially the membranes enclosing each pronucleus become broken at the point of contact and the nuclei embrace into a single structure which is then surrounded by a single continuous membrane. Within the membrane the chromosomes of male and female origin become arranged on the equator of the spindle: All of this may have taken an hour since the sperm bumped into the egg.

New biochemical activity, of which protein synthesis which we described earlier is but one example, now blows up in the fertilized egg. Details of how much of an increment there is in what enzyme or the diffusion of what ions is newly facilitated, are of interest only to the specialist. What is of overall import is that, from the balanced orchestration of these myriads of molecular activities, the new diploid cell in which the inheritable traits of the male and female parent are fused is readied to divide into two identical cells and thus launch into life an individual unique among all living creatures, including its own species and indeed its parents.

It is well known to biologists that in some species such as aphids the egg can develop into a normal individual without fertilization. Among bees and wasps, fertilization is elective: A fertilized egg develops into a female; the mere male arises from the neglected unfertilized eggs.

These are natural instances of virginal reproduction or natural parthenogenesis. It has been known for some 70 years that artificial parthenogenesis is also possible. Ripe sea urchin eggs have been exposed to a variety of chemicals limited apparently only by what was on the experimenter's shelf. Thus exposure to strychnine, chloroform, organic acids, toluene, ether, alcohol, and even high concentration of sucrose will start the development of sea urchin eggs. It is obvious from the wide diversity of these agents that they cannot cause the identical, unique reaction which initiates activation. Rather this unspecific diversity indicates that some factors are present and poised toward activation in the egg; the variety of often noxious external agents merely trigger this finely poised reaction. It may well be that the spermatozoan itself merely actuates this reaction. As evidence for this we recall that killed spermatozoa or indeed only spermatozoan parts such as tails can initiate the development of the egg.

Very interesting studies on artificial parthenogenesis in mammals were carried out by the late Gregory Pincus of the Worcester Foundation. These studies were merely incidental to this eminent physiologist's studies of the mechanism of hormone action, studies from which the feasibility of the use of the "pill" for controlled pregnancies developed.

Pincus induced artificial parthenogenesis in rabbits by two different manipulations. He removed rabbit eggs from unfertilized virgin does and incubated the eggs in nutrient medium for 48 hours. This shock was apparently sufficient to activate the eggs for, when they were returned into the fallopian tubules of foster virgin rabbit does which were made pseudo-pregnant either by mating with a sterile buck or by the injection of the appropriate hormones, several eggs resulting from such manipulation proceeded to develop and divide; indeed one gave rise to a live rabbit. In another series of experiments Dr. Pincus merely chilled the unfertilized eggs *in situ* within the fallopian tubules and permitted continuing development within the real mother by appropriate hormonal intervention. From such manipulations a stillborn and a live rabbit emerged.

Natural parthenogenesis is apparently possible even among birds. It was reported in a reliable study of domestic turkeys that 40 percent of the eggs laid by hens who had no contact with males developed and hatched into normal poults, all of which were females.

The question of the possibility of spontaneous parthenogenesis among humans insinuates itself into our minds. Claims of its occurrence have been made from time to time. Since none of them came from controlled situations, little, if any, credibility can be placed in them. With current technology such a claim could be easily verified or disproved. Since such a child would have the identical chromosome and, therefore, genetic endowment of the mother, pieces of skin from one could be grafted onto the other with total success.

The venerated Virgin Birth was, of course, also biologically possible, except that the determining force of the human sex chromosomes normally would have ordained a female child.

Thou cunning'st pattern of excelling nature.
SHAKESPEARE

Chapter four

Cell Fusion and Chromosomal Patterns

Just as Nature evolved an elaborate system for the fusion of sex cells, so it has developed a guard against the fusion of somatic cells. This is a system of elaborate membranes enveloping every somatic cell which permit some molecular communication but bar rigidly fusion of cells or the exchange of genetic material. Such restriction is, of course, an absolute need; otherwise, all of our somatic cells might fuse and instead of an exquisitely ordered system of cooperating organs we might become a huge blob of indiscriminately coalesced tissue growing in anarchic chaos.

However, no biological system is perfect. In the latter half of the nineteenth century keen pathologists would describe giant cells containing several nuclei. In 1873 a Swiss investigator, Lugenbühl, associated the existence of such cells with their proximity to small pox lesions in man.

It is, of course, a constant challenge to the biological scientist to repeat Nature in the laboratory. In this case it took some 90 years to repeat and even to surpass Nature. In 1965 the Oxford biologists H. Harris and J. F. Watkins managed to fuse somatic cells not only of the same species of animals but even of distantly related organisms such as mouse and man. To understand and to appreciate this achievement we must go back and admire the handiwork of a few determined men.

Bacteriologists have long been able to grow colonies of bacteria from single cells. The method was developed in the laboratory of the great German bacteriologist Koch, but the idea originated not with the great man but with the lady who washed his glassware. It is reputed that she suggested that the liquid culture media on which bacteria grow could be made semirigid by incorporating in it agar, a product from oriental seaweed. About 1 percent of agar incorporated in fluid which contains all the required nutriments for bacteria forms a semisolid, moist jelly which immobilizes a bacterium placed upon it.

When the bacterium divides, the two daughter cells stay in close proximity and so do the granddaughters. In 24 hours, the number of offspring originating from the original pioneer on the agar may number 10^8 cells which are packed together in a little mound about 2 millimeters wide and a millimeter high. The great value of culturing from single cells is that all of the bacteria in a given clone, having originated from a single ancestor, have the same genetic makeup. (Unless, of course, a mutation occurred in one of the progeny.)

Cajoling single mammalian cells to grow under similar conditions has been a long-held hope—or rather dream—of biologists. Successful culturing would provide homogeneous material impossible to obtain otherwise. Mammalian tissues are composed of several different cell types, often with widely differing structures and functions. If clones could be raised, as are bacteria, from single cells, then cells with identical genetic makeup and identical shapes and functions could become routine objects of study.

The goal is very easy to state, but the path to achieving it had been blocked by almost insurmountable obstacles. Mammalian cells are much more demanding in their nutritional requirements and in their ambience than are cells of bacteria. An *Escherichia coli* bacterium requires only sugar and salts to be incorporated into its agar medium. Mammalian cells are not that versatile in their synthetic capabilities. At least ten different amino acids are an absolute requirement since mammalian cells cannot synthesize them. Moreover, all of the long catalog of vitamins are essential. Thus a list of essential components for a culture medium for a mammalian cell is as long as the shopping list of the mother of a rich bride. Moreover, it is not enough merely to have the required ingredients. The relative amounts of various ingredients must be carefully determined. For example, one amino acid can inhibit the utilization of another amino acid similar to it if they are not present in appropriate relative concentrations.

A mammalian cell is usually part of a very complex organization. It lacks many abilities. When removed from a tissue, a cell will not survive for long. Some degradative enzymes which are held in check in the intact organism begin to break down the important macromolecules within the cell. Some of these macromolecules may break down spontaneously. The degradation through the intermediacy of the addition of a water molecule is favored energetically over the synthesis of the macromolecules. We spend a large amount of energy constantly throughout our lives both to synthesize macromolecules and to keep them intact.

A mammalian cell, which is 1000 times larger than a bacterial cell, is a heaven on earth for a bacterium that happens to alight upon it for it is rich in nutrients and water. Contamination of such seed cells to be used for culturing is an ever-present danger against which heroic measures must be taken. Fortunately, some powerful antibiotics such as penicillin are innocuous to mammalian cells—that is why penicillin is such a marvelous drug—and can keep cultures of mammalian cells sterile without interferring with their growth. Attempts to grow isolated mammalian tissues started as early as the first decade of the century. During the next two decades, despite the turmoil into which mankind was pushed, some small progress was made. The dominant figure of this period was a flamboyant French surgeon, Alexis Carrel, who was working at Rockefeller Institute. Carrel pioneered surgery of veins and arteries and went on to apply his virtuoso skills to organ transplants in dogs. Without an understanding of immune rejection, these exercises were meaningless, but Carrel would get the maximum of mileage (toward Stockholm) out of his transplants. He would have himself photographed with two beautiful greyhounds, black and white, whose left forelegs had been exchanged. The Swedes, for reasons best known to them, awarded Carrel their Prize in 1912.

At the end of World War II, several biologists set themselves the task of growing mammalian cells in tissue culture. One of the outstanding workers was Dr. Wilton R. Earle, who was working at the National Cancer Institute in Washington. With immense patience, Earle and his associates analyzed the many problems which they faced. For example, Earle was convinced that heavy-metal ions, such as iron or lead, even in trace amounts were deleterious to his cells during his attempts to grow them. To eliminate these ions, Earle devised an elaborate system of purification of all glassware. He had large vats of fuming sulfuric acid in which the glassware was boiled and then rinsed many times before being put into the next vat of fuming sulfuric acid, and so on.*

The patience and skill of the pioneer biologists were rewarded: Growth of mammalian cells in liquid culture became a reality.

---

*The frustrations of a compulsive perfectionist are many. Late one evening, Earle came back to his laboratory and saw the shadow of a man moving up and down in front of his sulfuric acid vats. It turned out to be one of the janitors who was dipping an object hanging from a chain into the sulfuric acid vat. To his horror, Earle, discovered that the man decided this would be an excellent way of cleaning his crucifix. A sign went up in front of each of those vats the next day: "Please! These sulfuric acid vats are not to be used for the cleaning of crucifixes."

With improved technology today we can grow mammalian cells in tissue culture media in practically any quantity; money is the only limiting factor.

These methods of culture taught us a great deal about the growth of mammalian cells. At first it was thought that cells once removed from the controlling forces of the intact organ of their origin would lose their characteristic properties and assume the shape and attributes of some primordial cell. But experience taught us otherwise. Cells continue to retain their characteristics for many generations. Indeed, they retain them so well that we can tell whether a skin cell came from a young or an old person: The number of divisions a cell line can undergo correlates well with the age of the donor. Cells started from a young donor can undergo as many as 50 divisions; cells from an older person are tired to begin with and peter out sooner.

The constancy of cells in tissue culture and their availability in massive amounts* make them ideal for many studies. Thus in assessing the effect of cancer-producing viruses, if the shape, appearance and crowding capability of cells are altered we can ascribe the changes to the influence of the infecting virus. But more about this in a later chapter.

Earle and his associate, Kathryn Sanford, had actually started to approach the much more difficult problem: trying to grow cultures from single mammalian cells. They had limited success by drawing up single cells into little capillaries filled with nutrient medium. Some of these cells divided and continued to grow.

At this point, Dr. Theodore Puck, of the University of Colorado Medical School, took on this challenge. Puck was trained originally as a biophysicist and his early career was spent studying bacteriophage. Feeling that that field was too crowded, he transferred his interest abruptly to mammalian cell biology.

---

*If all the cells yielded by 50 cell divisions from a young donor were nourished and kept alive, a mass of cells 2^{50} in number would be available. This equals 10^{15} cells and since an adult human has about 10^{14} cells the mass of cells grown in about 40 days would equal the mass of 10 adults.

This kind of switch requires the courage of a pioneer. The scientist must abandon a well-tilled field which yields dependable crops—in publications and grants—and strike out into the unknown, possibly hostile, unyielding wilderness. But the courage of the lonely scientist who tries to open a new field has its rewards. He does not have to read a mass of previous literature, some of which is often misleading. Moreover, since he is essentially alone he need fear no competition for priority of publication should he discover anything new and, equally important, he is not baffled by contradictory reports from some less than perfect competitors, the source of whose false claims it sometimes takes months to track down.

Puck tried growing colonies from single cells which he placed on semisolidified nutrient agar.* Growth was very poor and Puck decided that the nutrients were limiting. He hit upon an ingenious method of supplying lacking nutrients. He took a culture of cells grown in liquid medium and he irradiated it with x-rays of sufficient intensity to destroy their ability to reproduce. The DNA of the nucleus is particularly susceptible to such damage. However, such cells are still capable of performing many functions within their cytoplasm which are not dependent upon the integrity of DNA. Thus they can continue metabolizing and putting out into the surrounding medium many of their products.

*Puck gives an amusing anecdotal reason for switching to agar-solidified media for growing clones instead of the single-cell-in-a-capillary method of Earle. Puck and André Lwoff, the great microbiologist of the Pasteur Institute, were both visiting workers at the California Institute of Technology. For his monumental work on the latent viruses of bacteria, Lwoff had developed a method of picking up under a microscope single bacterial cells, under sterile conditions, which he could transfer into minute quantities of fresh nutrient medium in which the bacteria would begin to divide.

An elaborate machine, a so-called micromanipulator, was developed for the purpose. Essentially it contains a system of reducing gears so that movement of the operator's hand is reduced in dimension when it is transmitted to the capillary. Nevertheless, a very steady hand is required because slight vibrations can deflect the pipet from its target and destroy the experiment.

Puck asked Lwoff to teach him the technique. Lwoff, one of the great teachers of microbiology who has given generously of his knowledge to colleagues for half a century, readily agreed. Puck tried valiantly, but the results were less than satisfactory. The cells either failed to be sucked up or were dropped out prematurely. Lwoff finally terminated the lesson by saying, "In order to master this technique, Puck, you must abstain from strong alcoholic beverages for 20 years prior to attempting it." And presumably that is when Puck decided to abandon the capillary technique and turned to agar plates.

Puck called these partially incapacitated cells "feeder cells," which he laid on the agar first. On top of these feeder cells he placed a few isolated unirradiated cells. This simple expedient was spectacularly successful. Every single cell which had not been irradiated produced a distinct colony of homogeneous cells.

He repeated these experiments with cells from various organs, kidney, lung, skin, and from young and old animals. They all grew, each cell giving rise to a large number of cells which formed a clone. A clump of cells could be picked up daily and dispersed in suitable media and counted either under a microscope visually or by an appropriate instrument. From such studies we could calculate the minimum time it takes for a human cell to double. It turned out to be about 18 hours. This period, which is called the "generation time," appeared to be rather long compared, for example, to that required by bacteria whose generation time may be as short as 18 minutes. However, it must be recalled that a mammalian cell is about a thousand times larger than a bacterium; therefore, to produce all the material for doubling requires a longer time. Actually, mammalian cells manufacture their components considerably more rapidly than do bacteria.

With a generation time of 18 hours, a single human cell could multiply to the number of cells in a newborn baby in 30 days. Of course, there is a slight difference between a mass of cells and a human baby! A fertilized human egg multiplies very rapidly in the early stages of growth. If this rate were to continue during the whole period of gestation, the mass of the creature at the end of the nine months would equal the mass of the earth. Fortunately, some of the mechanisms of differentiation take a long time, and therefore only one Gargantua has been produced.

The patience and ingenuity of Puck produced a sharp tool for studying mammalian cells. It enabled us to bring the standards of quantitative science to the study of mammalian cell growth. Since 100 percent of the cells give rise to viable colonies we could readily determine the effect of various agents upon the growth of mammalian cells. Thus, for example, it was discovered that they are much more sensitive to changes in temperature than are bacterial cells. If the temperature is dropped from 100° to 86° F, or below, no cells will give rise to colonies, whereas bacterial cells will grow, albeit slowly, when their ambient temperature is reduced by 15° F.

The reason for this difference must lie in the greater complexity of the mammalian cell, which must rely for its growth and division on the products of a great many more enzymes. The effect of change in temperature on various enzymes is approximately, but not exactly, the same. As a general rule a drop of 10° C in temperature will reduce by one half the rate of reaction of chemical interactions. The cooling does not have an identical effect on all enzymes and, since a cell is shaped from the products of thousands of exquisitely orchestrated enzymes, growth and cell division cannot proceed in the cacophony produced by the cooling.

The newly devised methodology also enabled us to study the effects of such agents as x-ray upon human cells. The damaging nature of x-rays to human tissue became known some time after their discovery by Roentgen in 1895. Unfortunately, that information cost the limbs and lives of early workers in the field who permitted themselves to be exposed to x-rays without proper protective shielding. However, the susceptibility of human cells to x-ray damage could not be accurately determined until Puck's quantitative methodology emerged.

Before that we tried to infer the radiation sensitivity of human cells from data obtained from organisms which could be easily cultured. For example, the protozoan, *Paramecium*, resembles quite closely human cells in its structure. It was found that when *Paramecium* is exposed to x-rays it must absorb more than 20,000 roentgens before its capability to divide is destroyed. (A roentgen, abbreviated "r," is a unit of measure of radiation which produces ions. Thus an "r" produces a certain number of ions per cubic centimeter.) The findings of the resistance of *Paramecium* to x-rays were, unfortunately, much too reassuring.

There are tremendous variations among different organisms in their ability to withstand radiation. It is very dangerous and totally unwarranted to extrapolate the experience of one organism to the radiation resistance of another. There are great variations even among very closely related microorganisms. Some microbiologists who studied meat preservation by sterilization through irradiation found this out to their great surprise. They discovered that some organism was growing in meat which was exposed to sufficient radiation to kill any known microorganism. They isolated and studied these resilient bacteria, which they aptly christened *Micrococcus radiodurans*. Indeed, it has been found recently that *Micrococcus radiodurans* lives happily in atomic reactors!

Among mammals, too, there are large species variations in radiation sensitivity. For example, it was found in the Atomic Testing Ground in New Mexico that the kangaroo rat lives on the highly radioactive surface. Exhaustive examination confirmed that the kangaroo rat is indeed very resistant to radiation. It is reassuring to know that some organisms may survive an atomic holocaust. *Micrococcus radiodurans* and the kangaroo rat might start all over again!

After the methodology of growing human cells from single parents was perfected, it became relatively easy to study the radiation sensitivity of these cells. It is simple to plate 100 cells on several agar plates and expose each plate to a known dose of x-irradiation. Then in a few days we simply count those which were capable to giving rise to colonies. From such studies a startling conclusion emerged. The lethal dose, in other words, a dose of irradiation which prevents cell division in a human cell, is only about 80 r. This was an alarming finding in view of the reckless and indiscriminate exposure of people to x-ray for a variety of trivial and often nonsensical reasons.

Dermatologists had exposed the face of individuals to high doses of irradiation merely to try to remedy some trivial skin ailment. Moreover, shoe stores used x-ray machines to show the fit of the shoes that were being pushed on the customer. Mostly on the basis of Puck's finding, the allowable cumulative dose of radiation which is considered permissible has been greatly reduced. This is the reason why our dentists now suffocate us with the heavy lead shield on the rest of our body before exposing our teeth to x-ray photography.

Soon another bounty was to emerge from our ability to grow human cells with complete success. This was the visualization of the human chromosome. The complexity of the problem can best be appreciated when we realize that before 1956 the exact number of chromosomes in a human somatic cell was not known. The figure which was quoted inaccurately was 48. The reason for this was twofold. The chromosomes can be visualized as distinct structures only during certain periods of mitosis. Most of the time the chromosomes are simply frizzled coils within the nucleus. The cells for such studies had to come from some stillborn or aborted embryos. Moreover, the technology of preparing cells for visualization under the microscope was rather primitive. Thus different investigators reported different numbers of chromosomes, and an attempt was made to reconcile these disparate observations by the preposterous claim that the number of chromosomes in the human is variable.

An investigator, J. H. Tjio, reported in 1956 that the number of chromosomes in a human is probably not 48 as had been previously reported but 46. Dr. Puck invited Tjio to his laboratory in Colorado, and the two of them completed a classic study which not only gave us the correct number of chromosomes in the human cell but also defined the field so well that a new subscience of great practical importance emerged. (Incidentally, it also earned a well-deserved doctor's degree for Dr. Tjio.)

Their methods consist in arresting large numbers of cells in metaphase by the administration of a dye, colchicine, and then when such cells are placed in a salt solution whose concentration is lower than that which exists within, the cells swell and the chromosomes become separated and easily visualized. They examined the chromosomes of 2000 human cells. Of the 2000 cells counted, 99.9 percent revealed a chromosome number of 46 or of 92. The latter figure which is clearly the result of the doubling of chromosomes without cell division and is known as tetraploidy, had a frequency of less than 3 percent. Tetraploidy can result from cell fusion or incomplete division.

Puck and Tjio carefully measured and characterized each of the chromosomes and assembled them in a pattern, or an idiogram. Other workers designed similar but not identical idiograms. An international conference held in Denver resolved the discrepancies and drew up a final form of the idiogram or karyotype, which is now widely used. (Would that other international disagreements were equally reasonably resolved.) In Figure 4.1, the actual photograph of male human chromosomes is presented. The idiogram constructed from Figure 4.1 is shown in Figure 4.2. The chromosomes of a human female can be seen in Figure 4.3; they are arranged in an idiogram in Figure 4.4. The obvious difference is that the male has but one x chromosome and a small y chromosome. The female, in turn, has two x chromosomes. The x chromosome is one of the largest, and the extra chromosome in the female is three times larger than the y chromosome of the male, and thus the human female cell is approximately 4 percent greater than the male in chromosome volume. This larger genetic capacity may be the source of the well-known superiority of women.

The ease of manipulating human cells in culture and the improved technique of visualizing chromosomes permitted Puck to demonstrate exactly the effects of irradiation on human chromosomes. The high incident energy of the x-ray can break the chemical bonds between the components of DNA so that the strand actually breaks, and in turn the whole chromosome can be seen either to be distorted or to have a fragment missing from it, or these broken fragments can be seen to fuse together, producing monster chromosomes (see Figure 4.5).

In a whole organism such extensive irradiation would undoubtedly result in death, but the cells in tissue culture do not require all of the biological information needed by a whole animal. Thus, if the chromosomes which store the information for the development of the brain are damaged or distorted, there may be no deleterious effect on a cell in tissue culture; the cell can continue to grow with abnormal chromosomes which cannot be expressed in tissue culture under any circumstance.

A whole new area of investigation was opened up by these

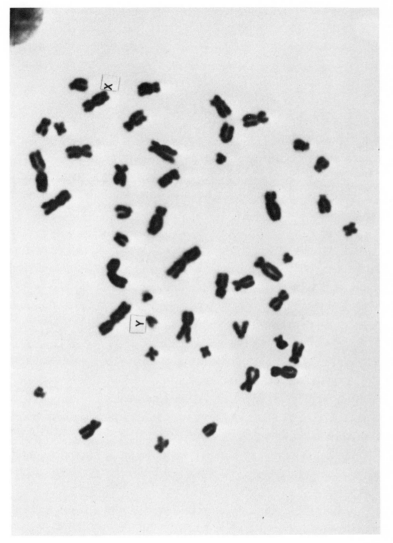

T. PUCK

Figure 4.1. Chromosomes of the Human Male

pathfinding observations. The number of genetic diseases to which man is heir is frightening. Some 800 different heritable deleterious traits have been catalogued which are dominants. In other words,

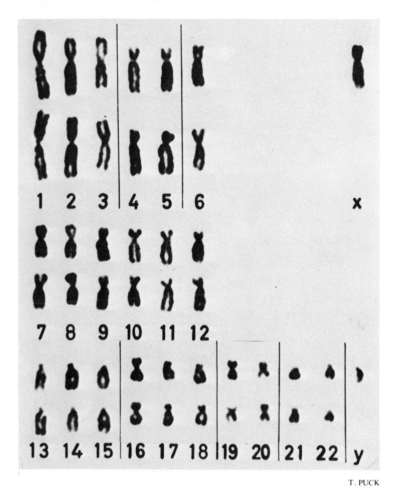

T. PUCK

Figure 4.2. Idiogram of the Human Male Chromosomes

a single gene mutation produces the harmful effect. Not all such traits are severely incapacitating, but a large number of them are. There are in addition some 600 known recessive characteristics where a double dose of the gene is necessary to produce the deleterious effect. As if these were not enough, about 120 deficiencies are known which are linked to the sex chromosomes.

Many of these defects are due to the absence or inactivity of an enzyme. At least 100 such metabolic errors have been catalogued

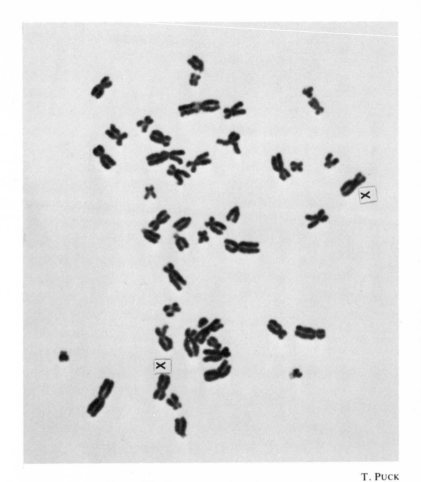

T. PUCK

Figure 4.3. Chromosomes of the Human Female

in man that are definitely due to enzyme deficiencies. The number of these defects is, of course, increasing or, rather, their identification is increasing with the improved biochemical technology available in some institutions. The defects are often detected by observing unusual products in an infant's blood or urine by highly sophisticated methodology. When a novel inborn error of metabolism is found, we can intercede sometimes; most often we cannot. Thus, if an infant is born without the ability to make urea, its prospects for survival are very very slim. Since the baby cannot eliminate two

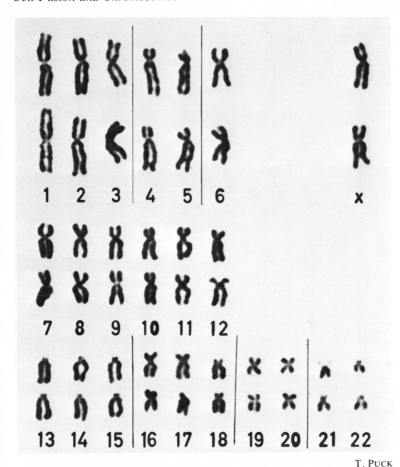

T. PUCK

Figure 4.4 Idiogram of the Human Female Chromosomes

toxic products, carbon dioxide and ammonia, via the innocuous molecule urea, the tissues of such an infant are laden with the toxic products.

Sometimes partial remedy of the enzymatic deficiency can be achieved, as in the disease known as galactosemia. Some infants are born lacking the enzyme to metabolize the sugar galactose, and this product piles up in the tissues of the infant. In low levels, galactose is a harmless, indeed, a useful component in a cell which, through the cell's enzyme-motivated alchemy, can be converted into energy; or from its atoms other products such as amino acids

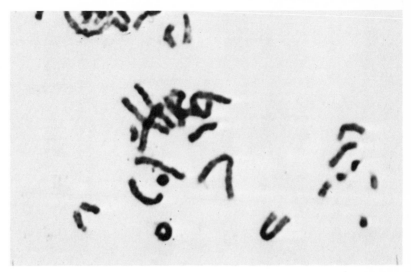

Figure 4.5. Chromosomes Damaged by X-rays   T. PUCK

might be fashioned. But living cells obey the ancient cliché about "too much of a good thing," and thus a large excess of almost any metabolite is toxic. The accumulating galactose damages the infant's liver and brain. Fortunately, the afflicted infant can be helped; galactose appears only in milk sugar in an infant's diet and, therefore, milk or any milk sugar products are not offered to that infant.

Sometimes the enzymatic deficiency, however, is not easily remedied. For example, in phenylketonuria, the enzyme deficiency cannot be so easily remedied and the child has to be put on a prohibitively expensive diet for about 10 years. In some cases, remedy is based on the addition of the product of the defective enzyme from some other source. For example, diabetes can be controlled by the injection of insulin obtained from cattle or, perhaps eventually, synthetically made insulin.

It is suspected that there is a deficiency disease corresponding to every enzyme present in a human. Since these enzymes may number in the tens of thousands, the number of possible syndromes is staggering. Fortunately, however, many embryos with such inborn errors of metabolism are never born; they simply abort early. However, such a large number survive that in 1967 it was estimated that 15 million people in the United States had congenital diseases affecting their daily life.

The obvious question is whether we can remedy such inherited lacuna in our genetic make up by restoring the absent or incomplete gene to the deficient individual. If we can transplant a heart, may we not be able to transplant a gene to its appropriate position on the chromosome? Such a feat can be done and is done routinely with bacteria. Certain bacteria which lack some gene may acquire the missing gene through infection by bacteriophage which do not kill them. This is not a transient effect since the acquired gene is still present after a great many cell divisions. The term "transduction" has been given to this phenomenon. The virus has to be grown first in an organism which is endowed with the gene in question. Then the bacterial virus may carry with it some of the DNA of the bacterium on which it grew. It is this fragment of DNA which, when it becomes integrated into the DNA of the recipient bacterium, restores the missing gene.

Could we develop a method of transduction of human sperm, or even of transduction of a whole human being with a gene which he lacks? What are the requirements? We suspect that every mammalian cell harbors a host of viruses which live in symbiosis with it. We shall discuss this in greater detail in a later chapter (on cancer). Would it be possible to grow a nonpathogenic virus in tissue culture on human cells which are endowed with the appropriate gene? Then, in turn, could we, with this virus, infect the sperm of a man who wishes to eliminate some of his genetic defects from his offspring?

Successful application of such an approach, which is called "genetic engineering," is in the distant future. In bacteria, we can restore the missing gene in one bacterium out of 10,000. In humans we are not interested in improving the genetic makeup of one in 10,000 individuals. We are interested in improving *one* individual.

Another type of deficiency disease is associated not with a particular enzyme but with chromosomal aberrations. They need not be inherited diseases; chromosomal aberration may occur because of some environmental factor during the period of gestation. For example, mongolism, or Down's Syndrome, occurs more frequently in infants born to women in their late thirties. This might be related somehow to some hormonal imbalance or the effects of aging on the ova. Such children are mentally retarded and have other pathologic alterations in many of their body organs. After karyotyping was perfected, it was discovered that an abnormal karyotype is invariably associated with Down's Syndrome.

In another type of chromosomal aberration males have an extra y chromosome; instead of the normal x-y chromosomal content they have x-y-y chromosomes. This chromosomal pattern is often associated with certain physical characteristics such as great height and skin highly disfigured by acne.

The subjects sometimes exhibit a sociopathic personality. For example, about 1 in 800 males is born with x-y-y type of chromosomal pattern. Among inmates of institutions for mental retardation or for mental diseases the x-y-y pattern occurs eighteen times as frequently as in the general population. The sociopathological tendency sometimes surfaces through hideous crimes.

Even larger multiples of the x chromosomes have been observed. Fortunately, most such fetuses abort spontaneously. They cannot survive the confusion of their chromosomes.

Still another well-characterized genetic deficiency is the one called xeroderma pigmentosum. These subjects, who may be normal otherwise, are extraordinarily sensitive to the ultraviolet rays of the sun. They develop large lesions in their skin which very frequently turn into cancer, and they die in their youth or early adulthood.

There is more than one deficiency in xeroderma pigmentosum. It is known that in some subjects an extraordinarily knowledgeable enzyme which can repair DNA is lacking. The discovery and study of this enzyme is a beautiful illustration of the value of all knowledge no matter how trivial and recondite it originally appears.

It has been known for a long time that the components of nucleic acids absorb ultraviolet irradiation. In other words, if we try to pass ultraviolet irradiation through a solution of a nucleic acid component, let us say thymine, we will not succeed; the solution is opaque to ultraviolet. But no change can be found in the thymine in the solution after such exposure. However, it occurred to a curious and systematic investigator to see what happens if ultraviolet irradiation is passed through a thymine solution which is frozen solid. He observed a profound change. Some single thymine molecules disappeared to form couplets or dimers of thymine. Thus the effect of the irradiation is different when the thymine is in solution or solidified.

It did not take very long for radiologists interested in biology to show that similar thymine dimers are present in the DNA of bacteria which are exposed to heavy ultraviolet irradiation. In due time, it was also discovered that bacteria contain a special enzyme which can remove, actually excise, such thymine dimers from DNA which had been exposed to ultraviolet irradiation. After the thymines are removed, appropriate enzymes reinsert single thymines guided by base pairing with adenine on the opposite DNA strand.

This is an extraordinary achievement of Evolution, protecting the organism's most precious material, the DNA, from damage caused by ultraviolet irradiation and restoring its integrity. Soon it was found that this enzyme, a so-called repair enzyme, is present in most human cells, but not in cells cultured from the skin of some patients with xeroderma pigmentosum. However, the syndrome is more complex than that since there are some subjects who do have this enzyme but, nevertheless, suffer the stigmata of this disease.

These diseases and abnormalities are a great financial and emotional burden on the families involved and, of course, they are a great burden on society as well. Fortunately, now something can be done about some of them. Diagnosis of the deficiencies of the embryo in the uterus is an increasingly applicable technique. Fetuses shed some cells into the amniotic fluid. Such cells can be recovered by a hypodermic needle as early as in the 14th and 16th weeks of gestation. These cells can be grown in tissue culture and their chromosomes can be readily visualized.

It is relatively simple to detect the chromosomal aberration in Down's Syndrome. Chromosome 21 instead of the usual double body has three bodies, as in Figure 4.6. If the embryo *in utero* is shown to be so afflicted, a therapeutic abortion can avert the birth of an incomplete human being and a lifetime of misery for the parents. Two percent of the women over the age of 40 become pregnant. In turn, they give birth to 30 percent of the children with Down's Syndrome. Therefore some experts believe that every pregnant woman over 35 should have an intrauterine examination—or amniocentesis—of the fetus she bears. The possible elimination of this human deficiency is within sight, provided some humane intelligence and skill are focused on it.

The sex of the fetus can also be readily determined simply by counting the x chromosomes. This is not done for some trivial reason such as predicting the sex of the child but rather to eliminate some sex-linked disease, such as hemophilia.

Some of the diseases which are inborn errors of metabolism because some enzyme is lacking can also be diagnosed *in utero*. The cells of the fetus can be grown in tissue culture and the enzyme suspected of being missing can be determined when enough cells have accumulated. If a pivotal enzyme whose absence is known to produce a severe abnormality in the newborn is lacking, the fetus can be aborted.

If the reader shudders at the thought of destroying a fetus simply because it lacks some enzyme out of thousands, let him consider the following. There is a condition known as the Lesch-Nyhan Disease which confers upon the afflicted mental deficiency and a tendency for self-mutilation. This condition can be so bad that the subjects will bite off their own fingers piecemeal and, of course, therefore must be kept in strait-jackets or completely tranquilized for the duration of their lives, in an institution.

A deficiency in an enzyme and a consequent flooding of the subject with uric acid has been found to be correlated with this

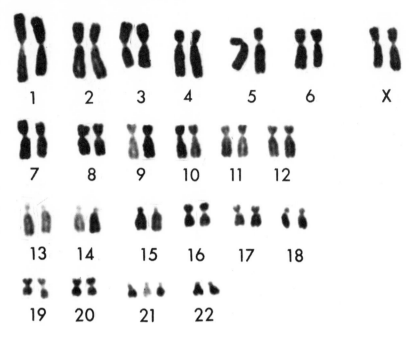

Figure 4.6. Idiogram of Down's Syndrome

T. PUCK

disease. It is not known whether the absence of the enzyme is coincidental or causal of the other dreadful symptoms. If a woman has such a child and yearns for a normal one—a very natural longing in such women—future pregnancies can be monitored via amniocentesis. The fetal cells are cultured and the presence or absence of this enzyme in the cells grown in tissue culture is determined. Thus the woman may be spared the agony of a recurrence of her first tragedy.

Our knowledge of human genetics has been synthesized into a powerful tool for the guidance of would-be parents. Genetic counseling is a new branch of medicine which flowered from all of this information. Prospective parents in whose lineage there is some genetic deficiency can be counseled on the probability of their having similarly afflicted offspring. Moreover, if the feared syndrome can be diagnosed *in utero* by either chromosomal or enzyme analysis, then the pregnancy can be monitored and, if need be, terminated.

Man's brain, that unique product of Evolution, may yet achieve its supreme goal: the improvement of the frail, disease-ridden body in which it is housed. This is no longer a Utopian dream; many techniques are at hand and new ones are constantly being perfected. Nor is the cost in wealth prohibitive. It would be probably less than one-tenth of the cost of the preparations aimed at destroying man, body, brains, and all.

After this long digression let us return to cell fusion. In self-defense I must point out that this digression was not imposed upon the reader by an undisciplined mind. Rather, it is the consequence of the pattern of growth of science: Science grows by digression. A highly imaginative or a keenly observant investigator presents some new phenomenon to the scientific world; if the phenomenon is significant, other investigators will seize upon it and apply it to their own work or shape it to fit their own creative pattern. If the recipient investigators are disciplined and their imagination is properly harnessed to the available experimental methodology, then the new frontier widens and yields harvests of practical benefit, such as genetic counseling or amniocentesis.

Of course, it is easy to throw off the traces which tie us to the currently available methodology and leap into realms of fantasy. For example, one can suggest off the top of one's head that some day we will be able to take cells from our skin and grow a mass of them in tissue culture equal in size approximately to our heart and, when the need may arise, we will switch on appropriate differentiation, inducing the mass of skin cells to be converted into a heart ready to replace our crumbling old one without the danger of immunological rejection. I consider this kind of imagination worthy of the unfortunate feeble-minded subjects I have discussed earlier. However, simpler goals are achievable. A piece of a patient's skin can be cultured in an apparatus which nourishes the cells and permits their multidirectional growth. It is astonishing that tissue grown in such organ culture retains its specificity of cell types. Such skin can be transplanted into the donor subject without fear of immune rejection. The method has been used successfully to cover and heal ulcerated wounds.

The fusion of cells from different animals into viable progeny cells by Harris and Watkins was a virtuoso achievement. Like all such achievements in experimental science, it was based on patient work of scores of investigators stretching back many years—in this case, indeed, half a century. Some attempts to culture tissues reach back to the first decade of the century.

As I said earlier, by the 1950s the methods of culturing cells in liquid culture had been sufficiently perfected that growth could be systematically achieved, and therefore the effect of various agents upon growth and cell structure could be studied. John Enders of Harvard was one of the pioneers in the study of the growth of cells in tissue culture. He was also one of the first to study, under controlled conditions, the effect of virus upon cells in tissue culture. Enders and one of his associates observed that the addition of measles virus to cells in tissue culture caused them to fuse, producing cells with several nuclei.*

However, the field was still murky because the fusion of cells was random and rare. The Japanese investigator Okada provided a highly useful tool which was to become a common reagent in a great many laboratories studying mamalian cells. Okada observed that a para-influenza virus which we shall abbreviate HVJ (hemagglutinating virus of Japan) is very effective in fusing animal tumor cells in tissue culture. He observed also that the virus is still effective as a fusion agent even after its infectivity is destroyed by ultraviolet irradiation. In other words, since the nucleic acid of the virus (in this case it is RNA) is damaged by ultraviolet irradiation, the virus cannot reproduce itself and thus destroy the cell, but it remains

*Enders received the Nobel prize for his work on the perfection of growth of kidney cells in tissue culture on a mass basis, because in turn the polio virus could be propagated in such cells and accumulated in massive quantities. The availability of the virus enabled Salk and Sabin to develop their methods of immunization.

active as a fusion agent. The term "Sendai" virus has been assigned to this virus, after the Japanese city where Okada works.* Technically, the achievement of fusion is quite simple. Sendai virus is grown, harvested, and exposed to a high dose of ultraviolet irradiation. After this molecular emasculation the virus is mixed with cells grown in tissue culture. The Sendai virus can invade the mammalian cell membrane via some unknown chemical reaction and, if the relative number of the cells is high, the same virus can invade the membranes of two adjacent mammalian cells. It thus forms a channel between those two cells. If enough of such channels are formed, the membranes between the two mammalian cells may disappear, forming a continuum between the cytoplasm of the two cells. In turn, eventually the nuclear membranes also fuse and a new enlarged nuclear membrane may be formed. (There is a schematic representation of this process in Figure 4.7.) Okada was very successful in fusing human tumor cells in tissue culture this way.

Harris and Watkins, in turn, chose a human cell line and a mouse cell line, both of them malignant, and brought them in contact with Sendai virus. They obtained fused cells some of which con-

---

* The quality of work in biological sciences originating from Japan these days is far superior to the scientific work prior to World War II. Before World War II much of the scientific work in the biological areas emanating from Japan was trivial and often not reproducible. However, the situation is completely changed now. The work is first-rate and scientific standards are very high.

This is a fascinating phenomenon which some sociologist ought to study. One of the reasons is the enlightened policy of the United States government in funding biological research in our country. It has enabled a great many American investigators to invite young Japanese scientists to come and work in our laboratories for a year or two. The effect has been twofold. It has enhanced our scientific productivity because there was a dearth of trained young Americans until recently.

The young Japanese Ph.D.'s have been fiercely industrious. One who worked with me was in the laboratory 15 hours a day 7 days a week. Finally, I insisted that he relax somewhat and I gave him a ticket to the ballet on a Saturday evening. On Monday I asked him how he spent his Saturday. "Worked until 7; went to ballet; after, returned to laboratory, worked," was his reply and he resumed his work at a furious pace. In turn, when these young investigators returned to Japan, they took with them the skills, the spirit, and the standards of American science. Thus, in a very short time, highly sophisticated and reliable research started to emerge from Japan.

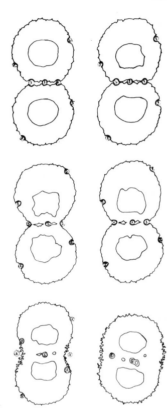

*Figure 4.7. Schematic Diagram of Fusion
by Sendai Virus*

tained two nuclei, some four. The origin of these nuclei could be readily ascertained both by their appearance and by labeling methods. (One of the cell lines may be grown prior to the fusing on a radioactive material which enters the nuclei preferentially and, when the two different cell lines are fused, only those nuclei will contain this radioactive label.)

Watkins and Harris showed that these fused cells can continue growing and, indeed, can go through many cell divisions. This is truly remarkable because there is a Babel of chromosomes in such cells. At first the number of chromosomes in a nucleus is the sum of the chromosomes of the two parent cells. This is a nightmarish condition since the chromosomes of man and mouse keep pouring out their information, causing chaos in the hapless cells.

But, as these cells with the supernumerary chromosomes continue to divide, they reject some of the chromosomes by not reproducing them; which chromosome is retained is easily seen from the characteristic shapes of chromosomes of different origin. The loss of chromosomes is unpredictable. It is humiliating to report that mouse chromosomes prevail over man's.

Cells with such a mélange of chromosomes are a versatile tool for a variety of studies. We can pinpoint the presence of the gene for a particular enzyme on specific chromosomes. Thus, if a certain chromosome disappears and along with it goes a particular enzyme which is normally present in humans, then the location of the gene for that particular enzyme is established. (The cells can continue living because the mouse chromosomes may be providing the needed enzyme. The enzymes from the two sources can be identified by chemical means because our cells retain their individuality even at the molecular level.)

Fused hybrids can be used for highly valuable studies of pathogenic mechanisms. For example, human cells in tissue culture are susceptible to infection by polio virus but mouse cells are not. The mouse, unlike man and other primates, lacks the entry sites in its cells for the polio virus; apparently, polio made its appearance late in evolutionary history. A human-mouse hybrid is susceptible to polio infection at first. Therefore we can study which human chromosomes house the information for the receptivity to polio because, as the chromosomes are lost, the infectivity may also disappear.

Processes of infection can be studied even better in other fused cells. When hamster and mouse cells are fused, both sets of chromosomes continue to be reproduced. A certain line of mouse cell permits the reproduction of a virus called polyoma but a hamster cell line does not. It is found that virus production is greatly lowered in such hybrid cells. Apparently, some product of the hamster chromosomes can inhibit virus production. The meaning of this for therapy of viral infections is obvious. However, an enormous amount of research to effect the concentration and purification of such factors will be needed before we can get any practical benefits from these observations.

Hybrid cells also lend themselves to studying the mechanisms which regulate the activity or, indeed, the absence or presence of an enzyme. We may cite a very interesting study in pigment formation as an example. Certain tumors of hamsters, so-called melanomas, produce a dark pigment even in tissue culture. Such a cell line has been fused with normal skin cells from a strain of mice which can produce pigments in the whole animal, but the cells in tissue culture cannot do so. The fused hybrids were colorless. The chromosomes of the hamster remained intact. Nevertheless, an important enzyme which is needed for the synthesis of the pigment was found to be lacking. Apparently, there is some product from the chromosomes of the mouse which prevents the formation of this enzyme in the hybrid cells. We therefore have a new tool for the study of the control of pigment formation.

Myriads of questions will be posed and answered with hybrid cells manipulated by men of ingenuity and skill, but they are beyond our current intellectual horizon.

Oh, the vanity of formulas.

APOLLINAIRE

Chapter five

Cellular Orchestration: Regulation

The most fascinating period of an orchestral concert, to at least one eccentric concert goer, is the few minutes before the entry of the conductor. Every one of the perhaps 100 musicians is letting go: Some practice a difficult passage, some just blast away, delighted to be free of the constraints soon to be imposed upon them. Soon there is silence, and then the baton signals and the 100 possible sounds are turned on and off, yielding a rounded, enthralling mélange of sounds. If we expand the participants a hundred fold to encompass the perhaps 10,000 enzymes in a mammalian cell, we have a condition resembling the orchestra before the arrival of the conductor. Such a cell has been called "toti potent," containing as it does all the biological knowledge with which Evolution has endowed

97

the species of that cell. But its hoard of knowledge is practically never expressed totally and simultaneously.

There are cues and signals which turn on and off this potential information. The need for regulation of the expression of all of our information is twofold. The development of our ultimate size, shape, and variety of tissues must follow a precise blueprint. The bacteria, such as the colon bacillus we have discussed earlier, have never learned to do this. They just keep growing and multiplying, by cell division, as long as their food lasts or until they befoul the medium where they live with poisonous products which kill them. Given adequate food and methods for neutralizing the acids they produce, bacteria can grow indefinitely until tons of them accumulate. There are some industrial processes for which growth of bacteria for a variety of purposes in such amounts is routine. Thus the enzymes which are incorporated in some washing preparations are extracted from bacteria grown on such massive scales. Probably never before in the history of the universe have these bacteria accumulated in any one place in such numbers.

The second, equally important, need for regulation is to maintain a variety of products within the cell at levels which are salubrious to the cell or to the organism. In excess, everything is toxic. A simple example of this is what is facetiously known as the Chinese Restaurant Syndrome. The commercial soy sauce these days is essentially glutamic acid, one of the amino acids in our diet and in our body. Certain bacteria produce glutamic acid in large amounts. The Japanese food industry purifies this glutamic acid, neutralizes it partially with alkali, adds caramel for soy coloring, and ships it out with exotic Chinese trade names. The sauce thus prepared has never been a component of a soy bean plant. An excess of glutamic acid, when injected into animals, produces severe seizures. When the acid is eaten in large amounts, violent headaches may ensue in the sinophile customer. This is an example of the accumulation of a metabolic product from external sources—from overeating.

But glutamic acid is readily synthesized in our body from simpler precursors. How can we prevent excessive synthesis so that it will not accumulate in toxic amounts? The method of control was obscure until 1941 when a remarkable biological scientist showed that there are feedback controls in our cells to regulate levels of metabolites.

Feedback has been defined by Norbert Wiener, the MIT mathematician, as "a method of controlling a system by reinserting into it the results of its past performance." Feedback machinery has been used by man ever since antiquity. Thus a very ingenious mechanic of the Ptolemies, who apparently lived about 250 B.C., designed a clock which was operated by a steady flow of water. The height of the water accumulating was controlled by a float which in turn activated a valve to shut off the flow of the water. This is, of course, the system that is used today in maintaining the level of water in a water closet.

In modern times, a very important feedback mechanism was that devised by Watts, the inventor of the steam engine. This is a device called the governor. It consists of two arms with two heavy spheres attached to their ends. As the steam engine increases its speed, as the result of the increased pressure of the steam, the two arms rise by centrifugal force, and in so doing they activate a valve which permits some of the steam to escape. As the steam escapes, speed is reduced and the balls move downward, thus shutting off the escape valve for the steam. This centrifugal "governor" is the origin of the term "cybernetic" which was coined by Norbert Wiener for feedback systems.

Evolution, that supreme inventor, had devised feedback control, or cybernetic control, eons ago, but the mechanism lay hidden until 1941 when it was revealed to the searching eyes and mind of an unusual scientist. Zacharias Dische is a Polish physician who never practiced medicine but turned to research instead.

Dische was a pioneer in the study of the pathway of metabolism of glucose. From the patient research of many investigators, including Dische, we know that the energy in glucose is not released in one single burst. There are a series of enzyme reactions which modify glucose and some subsequent products only slightly. Thus, since the energy is not released in one single burst it is released in a controlled manner. The first step in the metabolism of glucose is the addition of a phosphate group to form a product called glucose

6-phosphate. (There is a phosphate on the 6th carbon of glucose.) Four enzymatic reactions later, a product which contains only 3 carbons but 2 phosphates, diphosphoglycerate (DPG) is produced.

At the start of World War II, Dische had to flee from the Nazis who were engulfing Europe. He worked in Paris and later in Marseilles for a laboratory which was developing techniques for blood transfusions for the French Army. Dische spent his days at tedious routine, but at night he sneaked back to the laboratory, appropriated some blood, and continued his investigations of the biochemistry of glucose metabolism. He then discovered that, in the intact red cell, the accumulated DPG shuts off the first enzyme which makes glucose 6-phosphate. He understood immediately the meaning of his discovery.

The end product, DPG, interferes with an unrelated enzyme reaction several steps away, and in this manner DPG controls its own concentration because the first step leading to its eventual formation is shut off. Dische escaped to America in a coal barge and was given a haven at the Biochemistry Department of Columbia. He continued to make first-rate contributions and still does though he is well past 70. He did absolutely no self-advertising, wrote no elaborate review articles about his many contributions. He has received no honors, but he is a happy man. He counts himself among the privileged of mankind because for the past three decades he had the facility and the freedom to do what he wanted: to pursue his ideas in the laboratory. It is good to know that there are a few scientists who seek as their only reward the search for, and occasional discovery of, truths hidden by Nature from the eyes of man since the beginning of time.

Feedback regulation of the formation of metabolites was rediscovered fifteen years later, and it was found that it is a very frequently used mechanism, especially in bacteria. The beauty of this system of control is obvious. The synthesis not only of the final product but also of all of the intermediate products is also shut off. Thus in one fell swoop the cell is protected against the accumulation of perhaps four or five products which may be toxic to it.

The next question that was raised is obvious. How can a single protein, a single enzyme, be that knowledgeable? It must recognize the primary substances, glucose and phosphate, and it also must recognize that an end product five steps away is increasing in concentration beyond tolerable levels. The answer which emerged from patient investigation is a fascinating one. Enzymes which are under the control of feedback mechanisms are not simple proteins; rather, they are combinations of two or more proteins. One of these is the original enzyme; the other part is a regulatory protein which recognizes the distant end product. Upon recognition, the regulatory protein and the end product combine and in a concerted way limit the catalytic ability of the enzyme part of the complex.

Product inhibition of chemical reactions has been known for a long time. Every beginning student of chemistry is badgered to memorize Le Chatelier's Principle which states essentially that in a reversible reaction, such as that between nitrogen and hydrogen to produce ammonia, the accumulation of the end product, ammonia, can reverse the reaction to yield nitrogen and hydrogen again. Le Chatelier's Principle, thus, is a specialized case of Newton's Law of Inertia: "To every action there is a reaction." However, what Evolution has discovered in feedback regulation by the end product of a series of reactions is sheer genius. The end product shuts off not only the reaction which produced it but also shuts off a whole pathway of as many as a half a dozen or more sequential reactions.

Feedback inhibition operates only to put brakes on the activity of preexisting enzymes. It does not control the synthesis of the enzyme from the information in the genetic material of the organism. There are mechanisms for the prevention of the *synthesis* of an enzyme which either is not needed or its product has accumulated to a sufficient extent. Such regulation exists in many bacteria where it provides a great advantage. The tiny bacterial cell need not be burdened with the constant presence of all of the enzymes it is capable of synthesizing.

This finding is easy to state, but it took half a century to establish definitively the existence of such mechanisms. It was first observed around 1900 by a bacteriologist, Dienert, that certain microorganisms can adapt to grow on specialized nutrients. Their adaptation was slow but definite. Not much was known about enzymes in those days, and Dienert ascribed the phenomenon to a Darwinian selection of organisms which preexist and which can utilize the specialized nutrient and thus outgrow other, less well-endowed organisms.

Not until 1930 was it definitely shown that bacteria can indeed elaborate a specialized enzyme to utilize some special nutrient. This was reported by an investigator, Karstrom, as part of a Ph.D. dissertation at Helsinki, Finland (one of the rare instances of real originality in such an exercise). He grew some *Escherichia coli* on sucrose, or cane sugar, and another batch of it on lactose, the milk sugar. Both of these sugars have as half of their constituents the glucose molecule. He harvested the *E. coli* grown on the two different sugars and found that both batches could ferment glucose, but only those which were grown on lactose or sucrose could ferment lactose or sucrose, respectively. He called certain enzymes constitutive, such as the ones which ferment glucose, because they are always there, and the enzymes which ferment lactose or sucrose he called adaptive enzymes.

However, the suspicion still lingered that perhaps there is a Darwinian selection of preexisting cells. Marjorie Stephenson, a gifted English microbiologist, laid this objection to rest when she showed that, during the period of adaptation, a period during which the lactose-splitting enzyme is emerging, the number of bacterial cells does not change. That is where the problem stood at the start of World War II.

After World War II a brilliant Frenchman, Jacques Monod, who was working at the Pasteur Institute in Paris, took up the problem.*

---

*Monod is one of the few "Renaissance" men of science. Small in stature and with one of his legs damaged by polio, he nevertheless became one of the great mountain climbers of Europe. Such physical prowess served him well during World War II when he was

In probing the mystery of these enzymes which can be produced in response to a need, Monod had many collaborators, most of them Americans who made pilgrimages to the Pasteur Institute, which after the war was an outstanding center of experimental biology. But the most brilliant contributor to the problem was another worker at the Pasteur Institute, François Jacob.*

Jacob and Monod pooled all the information known about adaptive enzymes, both chemical and genetic, and came up with a very attractive hypothesis on the mechanism of functioning of adaptive, or, as they renamed them, inducible enzymes. The Jacob-Monod hypothesis is so attractive and relatively simple to grasp that it has become a household word among biologists and has penetrated even biology classes in high schools, where most of the young scholars think of Jacob-Monod as one man.

According to the Jacob-Monod hypothesis, synthesis of an inducible enzyme is under complex control by several genes and by some factors in the environment. In the first place, it is postulated that the information for the synthesis of an enzyme which will cleave lactose into its two component units is inscribed in a gene called a structural gene. Close to the structural gene on the chromosome of the bacteria are two other genes, the regulator gene and the operator gene. The whole complex is known as the ''operon.''

he head of intelligence for the Paris underground and made trips over the Alps in the height of winter to Switzerland to keep Americans informed, exploits for which he was decorated by both the French and Americans.

He is a cellist of solo concert rank but he switched to biology, the field in which he received his Ph.D. As the growing intricacies of the adaptive enzymes required it, he became a mathematician and molecular biologist. When he turned to being an essayist he succeeded equally. His ''Chance and Necessity'' is a global best seller.

*At the start of the war, Jacob escaped from France to England where he joined De Gaulle's forces. He was one of those dauntless Frenchmen who marched with General Leclerc from Lake Chad up to northern Africa. Jacob also fought with the French division in the Normandy invasion, where he was badly shot up and invalided out of the Army. After military service he studied medicine but became disenchanted with that profession and went to the Pasteur Institute to take a doctoral degree in science. Soon he became one of the most ingenious bacterial geneticists.

The operator gene is essentially a switch for the structural gene. When the switch is closed, the structural gene cannot be transcribed by the DNA-dependent RNA polymerase which normally would produce that particular messenger RNA. The machinery of the switch on the operator gene is a protein which binds to it, and as long as it is bound the switch remains closed. This protein is made by the regulator gene.

When an inducer such as lactose accumulates within a bacterial cell, it combines with the repressor protein so that the protein binds less effectively to the operator. It thus drags the switch open and the polymerase enzyme can start the transcription of the messenger RNA from the structural gene. This hypothesis has found confirmation in some systems; the repressor protein which binds part of the lactose operon has actually been isolated, and we could actually show *in vitro* that it binds to DNA.

The picture became complicated when it was discovered that induction could occur not only from the *presence* of some substance but also in some cases by the *absence* of a metabolite. Thus, for example, if *E. coli* is taken from a culture medium which is rich in amino acids, several enzymes for the synthesis of some of these amino acids will be lacking from the organism. However, if the organism is placed in a medium which lacks some of these amino acids, the enzymes for their synthesis slowly appear.

The two brilliant Frenchmen proposed for this a somewhat different model, namely a repressor on the operator gene which is bound tightly to the operator when the product of the enzyme is present, so that the switch is closed. As the product is consumed, the switch is opened by a less tight binding of the repressor to the operator or the bacteria are said to be derepressed. This part of the predicted mechanism was only partially confirmed by the emerging facts. In the case of the operon for the synthesis of some amino acids, the switch is not a protein but the transfer RNA for that amino acid. It comes to the appropriate operator gene—how it recognizes it is a mystery—and modulates the synthesis of that particular amino acid, so the transfer RNA can become loaded and return with its amino acid to the ribosome and participate in protein synthesis.

Jacob and Monod postulated all such regulations by assuming that the messenger RNA is unstable and after a while it decays and, therefore, new messengers have to be produced by the operon when it is derepressed or induced. The conclusion by Jacob and Monod about the transient nature of messenger RNA came from studies of very specialized systems. The American biochemist, Volkin, observed that after infection of *E. coli* by a phage a new RNA appears which mimics in composition not the DNA of the *E. coli,* but that of the phage. This RNA *is* rather unstable.

As so often happens in science, a brilliant hypothesis may be so persuasive and may become so widely accepted that it inhibits investigations in other directions and, indeed, may inhibit the introduction of new ideas. This is precisely what happened after the acceptance of the Jacob-Monod hypothesis by the scientific community. However, it soon became evident that this cannot be the only control mechanism. In the first place, even in bacteria not all of the enzyme conglomerates or enzyme clusters are under the control of an operon.

It is highly doubtful that the operon model exists in mammalian cells or indeed even in yeast cells. Moreover, the key tenet of the control by the operon is that the messenger RNA becomes unstable and decays after a while. Studies with mammalian cells, for example the red cells, have shown that the mammalian messenger is not unstable. Mammalian red cells continue making hemoglobin after the nucleus with its information-bearing DNA disappears. Therefore no new messenger RNA can be synthesized.

Still another line of evidence for the stability of messenger RNA comes from an unusual aquatic plant, acetabularia, which thrives in the Mediterranean Sea. It was introduced into research by the Belgian biologist Brachet who likes to combine scuba diving with his scientific pursuits. Acetabularia is a green plant which thrives on the floor deep in the sea. It may be an inch long, but it is a *single cell* whose nucleus is confined to one end. At the other end of the stalk there is a sort of cap, an organ whose surface is large, the better to absorb the feeble light reaching it. Brachet showed that the nucleus of acetabularia can be removed (a surgical procedure of infinite ease; one just snips it off with scissors), and the plant will continue to grow, forming its characteristic cap. Therefore the messenger RNA which had been formed in the nucleus before the snipping of it must continue to be accurately active.

We can transplant nuclei from acetabularia of one cap type to those of another cap type. At first there is confusion in the formation of the cap; a hybrid type begins to appear, but after a while the cap type of the newly transplanted nucleus takes over.

There is no evidence that the messenger RNAs in animal or plant cells decay rapidly, and it is known now that, although mammalian cells apparently transcribe all of their DNA, little of that information is translated into proteins. Consequently, other mechanisms of regulation had to be entertained. One is that not all of the RNA leaves the nucleus; the second is that there is selective translation of some of the messenger RNAs during protein synthesis. At the present time, evidence indicates that probably both of these mechanisms operate.

The evidence for selective control of translation is becoming increasingly compelling. The mechanism of translation is the most complex of all biological processes. It has more obligatory components than any other system known. Consider its complexity. First of all, there are the ribosomes. They are relatively huge spheres containing RNA and as many as 30 different proteins. The next component, the messenger RNA, is the simplest. It consists of a long chain of bases which reflect, by complementarity, the bases in DNA. Messenger RNA is essentially a long tape.

The next component, transfer RNA, as stated earlier, is the most complex macromolecule found in Nature. It is also the molecule which is the most versatile in function. Transfer RNAs are the molecular vehicles to which appropriate amino acids are attached, and they carry the amino acids to the ribosome, where protein is assembled from the amino acids according to a sequence which is encoded in the messenger RNA. Therefore transfer RNA must have the ability to recognize the message as well as the appropriate amino acid. It is thus the bilingual interpreter of two different sets of symbols: those of the structure of each of the three nucleic acid bases and of the diverse structures of the amino acids. For these tasks, transfer RNA has been endowed with an extraordinarily complex structure. It contains some 80 of the same bases from which are made the other nucleic acids: adenine, guanine, cytosine, uracil. But, in addition, it contains some 50 different modifications of these bases.

The origin of these modifications was obscure until some 10 years ago. Then, through a serendipitous observation and some hard work, it was demonstrated clearly that these modifications do not occur during the synthesis of the primary sequence of the 80 bases, but certain knowledgeable enzymes come and modify these bases after the whole nucleic acid is woven together. The set of enzymes discovered first are those that modify the bases in transfer RNA by inserting a special configuration of atoms called a methyl group. This is composed of a carbon atom with three hydrogen atoms attached to it. The methyl group, though small, is rather bulky in three dimensions and confers large changes on the shape of transfer RNA. The existence of these enzymes pointed the way to search for other modifying enzymes, so that now some 50 of them are known. These enzymes turned out to be specific for species and indeed for organs; in other words, the enzymes of the heart are different from the enzymes in the liver.

These discoveries, which have been amply confirmed, were made by two very able students of mine and there has been no controversy about them. However, I must caution the reader that, as far as these findings relate to regulation, they have not received 100 percent acceptance, and in telling the story I may not be the objective historian. I suggested that these modifications, which are species- and organ-specific, must have a function that is organ- or species-specific. The most obvious of such functions is, of course, the production of organ-specific proteins during differentiation. It was found after 8 years of research that these enzymes which alter the structure of transfer RNAs do indeed change in every biological system undergoing changes in differentiation, in particular in can-- cerous tissue. We shall discuss these in Chapter Seven.

The enzymes which modify the structure of transfer RNA are peripheral to the translation mechanism. But there are still other peripheral enzymes: the "charging" enzymes which recognize the particular amino acid to be placed on the appropriate transfer RNA. This is an extraordinary recognition system whose error is less than one part in ten thousand. It *must be* that accurate; otherwise, a wrong amino acid might be woven into the protein and we know that this can be disastrous. For example, in sickle cell anemia it is only one misplacement of an amino acid that makes the difference between the lethal pathological hemoglobin and the normal hemoglo- bin.

Marked changes in the charging enzymes have been observed in a number of systems undergoing changes in differentiation. These changes may confer new properties on the enzymes so that they will recognize only certain transfer RNAs which themselves had been changed and are ready for regulatory functions. However, this field of research is still in its infancy because it presents heroic challenges to the investigator. Transfer RNA is a macromolecule the determination of whose precise structure is difficult and tedious and, in turn, the charging enzymes are macromolecular proteins and the determination of *their* structure is equally difficult and tedi- ous. Only the most determined and dedicated investigators have stayed and continued to stay in this field.

For example, Bruce Ames, a persistent biochemist, proposed some 10 years ago that in a certain microorganism, *Salmonella,* the synthesis of the amino acid histidine is under the control of an operon different from the Jacob-Monod model. Ames postulated that the enzymes which synthesize histidine—and nine are needed for nine sequential steps—are controlled by the state of loading of the histidine transfer RNA. When the latter is fully loaded, it somehow shuts off the production of the enzymes. When the histidine transfer RNA is empty, it signals the start of the synthesis of the enzymes to produce histidine.

Ames found a mutant of *Salmonella* in which the enzymes of histidine synthesis are always turned on; they need not be induced; they are always derepressed. How does this mutant differ from the usual *Salmonella*? Ames showed with great skill and patience that the histidine transfer RNA in the derepressed mutant was structurally different from that found in the "wild type" *Salmonella*. The difference is a miniscule one; two of the 80 bases had not undergone a minor modification by the appropriate modifying enzymes. How these minor changes affect the interaction of the transfer RNA with other macromolecules we do not yet know. Our knowledge of the physical chemistry of macromolecular interactions is too primitive for the task. Our biological knowledge has surpassed our growth in chemical skills.

Any of the many components of the translation process could be regulatory factors for the translation mechanism. If any of these components is missing or is defective, protein synthesis must stop. Of course, this would be too crude a system to consider a regulatory one, because it is not selective: All protein synthesis would have to stop when a component of the system is lacking. The most likely candidates for a selective regulatory function in translation are the transfer RNAs because of their wondrously complex structure. One of the earliest pieces of evidence of control at translation clearly showed transfer RNA to be the regulatory factor. These are the transfer RNAs to which the unfortunate term "suppressor transfer RNAs" has been given. "Permissive transfer RNA" would be much more appropriate.

We may summarize such a system in this way. Mutations can occur so that the DNA, when copied, provides not the appropriate triplet for an amino acid but, by mistake, a terminating signal triplet in the middle of a message. Protein synthesis cannot proceed beyond that point because the protein is prematurely terminated. There are known instances where through an extraordinary coincidence there is a mutation in the transfer RNA itself so that it coincides with and complements the premature terminating codon in the messenger RNA, thus permitting the filling of the gap with an amino acid so that protein synthesis continues. It is an extraordinary coincidence to have a mutation on the transfer RNA complement the mutation on the DNA and, consequently, the messenger RNA. Such a mutation really cannot be considered a regulatory mechanism because it occurs too infrequently. Nevertheless, it serves as a model for other translational controls.

We may visualize a more generalized control of the synthesis of a given protein by transfer RNA through the absence of a specific transfer RNA which is required at a certain point in the synthesis of a given protein. Such a subtle mechanism is very difficult to demonstrate in a complex *in vivo* system. The success or failure of a demonstration or, indeed, of the biological scientist himself, depends upon the choice of an appropriate biological system which is amenable to manipulation to achieve a certain demonstration.

A Canadian husband and wife team, Wainwright and Wainwright, has successfully selected such a system. Hemoglobin synthesis in the chick embryo is programmed to commence at a definite time. One can remove early chick embryos by means of scissors and place them on a filter paper in appropriate nutrient medium and incubate them at a salubrious temperature. The development of the chick continues, and at the specified time hemoglobin synthesis commences. One does not even need a sophisticated methodology to determine this. It can be seen visually that the developing embryo is producing hemoglobin: It turns red. The Wainwrights isolated transfer RNAs from a late stage of the embryo where hemoglobin synthesis was already turned on. They added such a mixture of transfer RNAs to the premature embryos which were not making hemoglobin. And lo! hemoglobin synthesis was switched on. They could track down by patient analysis that only one of the transfer RNAs is needed for switching on the premature hemoglobin synthesis. Therefore the messenger RNA for the synthesis of hemoglobin must preexist in the primitive embryo; it merely awaits switching on for translation by an appropriate transfer RNA.

Another curious example of translational control of protein synthesis has emerged from studies of aging wheat leaves. The tip, or the apical region, of a wheat leaf stops growing when a certain size is reached. Investigators at Oak Ridge Biological Laboratories have collected enough of the apical portions of wheat leaves to isolate from them transfer RNAs. One of these transfer RNAs can still accept its appropriate amino acid, but cannot attach itself to ribosomes. It is now known that a modification next to the code in transfer RNA is essential for attachment to the ribosomes. The nonfunctional transfer RNA in the wheat leaf was found to be incompletely modified at that site. To keep young we may have to look to our transfer RNA modifying enzymes!

Our understanding of regulatory mechanisms is so fragmentary because Nature and Evolution together are such very ingenious inventors, developing mechanisms which we cannot predict but can only decode by very patient and persistent studies. A good example is our current understanding of the mechanism of control of RNA synthesis in bacteria. It was noticed over 20 years ago that, if an amino acid which is essential for the growth of the colon bacillus is withheld from the organism, the synthesis of RNA is very rapidly shut off. This, of course, makes good sense. If there is a missing amino acid, protein synthesis cannot be completed and, if there is to be no protein synthesis, there is no sense wasting energy in synthesizing RNA whose main function is the synthesis of protein. But the mechanism was baffling.

How is absence of an amino acid sensed and interpreted by an organism as a signal to shut off synthesis of RNA, which contains no amino acid? There is an obvious barrier to converting the language of the amino acids into the language of the bases composing RNA. Moreover, there is a geographical barrier between synthesis of proteins and synthesis of RNA. The former takes place on ribosomes which are all over, the latter is restricted to the DNA of the cell. A chance observation revealed by accurate determinations made 18 years ago by one of my assistants put us on the trail of the mechanism of this control. We discovered a species of *E. coli* which was *unable* to control its RNA synthesis in the absence of an amino acid. It continued making RNA. The term "stringent controlled'' was given to the large majority of *E. coli*, which were able to turn off RNA synthesis, and "relaxed controlled" to those that were unable to shut off RNA synthesis.

Other investigators showed that stringency and relaxedness are genetically inherited traits, and the hunt was on for the factor which the genes of the relaxed organism could not make and those of the stringent ones could. After 18 years and hundreds of publications by colleagues all over the world, we now see the mechanism of this control. During the idling of the machinery of protein synthesis for lack of an essential amino acid, in the organisms with stringent control the ribosomes put out a product which is a modified derivative of guanine, one of the components of RNA. When this product accumulates in such a cell it inhibits the enzyme which copies DNA into RNA. The organism which lost control over RNA synthesis is unable to make this modified guanine. We still do not know whether this particular regulation operates in animal cells.

Multicellular organisms have evolved still another method of control. They can elaborate hormones (from the Greek *hormon*, "to stir up"). These are substances secreted by some cells into the blood stream to affect the activity of other cells. In brief, then, hormones are messengers which are obeyed implicitly. For example, if an animal is frightened, the hormone adrenalin will pour out into its bloodstream. The hormone will relay the message of fear to cells in the liver which will promptly begin to break down glycogen, the storage molecule for glucose, pouring out glucose into the bloodstream to provide energy to the various tissues and muscles to flee the danger or to fight.

The mechanism of hormone action was mysterious until some 25 years ago. It was presumed by physiologists that we would not be able to deduce the action of hormones at the molecular level because it was believed that the action of hormones could only be studied in the whole organism. The complexity of such a study would bar an approach to studies at the molecular level. This kind of holistic pessimism was a paralyzine brake upon progress in many fields. Physiologists used to compare those of us who were striving to study life at the molecular level to watchmakers who would grind up a watch in a mortar to find out how it works by isolating and studying the dismembered debris. One wonders how these people received the news when Watson and Crick deduced not only the structure of DNA from mere x-ray pictures of isolated purified DNA but more importantly the transfer of biological information by chemical base complementarity. Such information was thought to be forever hidden from us since it was supposed to be enshrined in the arcanum of the total organism.

After World War II, two factors contributed to the release of this paralyzing brake. A new generation of biochemists and physiologists who had been inspired by the rapid pace of progress in studying molecular mechanisms entered the study of hormonal activity. Moreover, hormones, which act in miniscule concentrations and therefore were difficult to study were made in radioactive form. This permitted following the pathway of minute traces of hormones when injected into tissues even though the dilution was enormous.

First of all, it was shown that hormones do not react randomly with all tissues and all organs. They are specifically concentrated in the so-called target organs. For example, it is known that the target organ of the female sex hormone, estradiol, is the uterus. This can be demonstrated in either of two ways. We can inject the hormone estradiol into immature female experimental animals, whose uteri are not developed, and within hours visible changes are observed in the uterus only.

The same effect can be demonstrated in adult female animals whose ovaries are removed experimentally. In these animals the uterus atrophies, it becomes small, immature looking, and there are changes in the cell types composing the organ. Within hours after the injection of minute doses of estradiol into the animal the uterus begins to regenerate. With the aid of radioactive label in the hormone we are able to show that in the target organ, the uterus, there are special proteins, receptor proteins, which accumulate concentrations of the hormone hundreds of times higher than in the rest of the animal's tissues.

But how is the message of the hormone translated within the target cells into the variety of new activities that follows the injection of the hormone? The answer to this question came from the work of a perfectionist research-physician, Dr. Earl W. Sutherland. Sutherland had been engaged in research before World War II; after his return from the war, where he was a medical officer, he faced the dilemma of many young physicians: which path to follow, medical practice or research? The former tends to be more certain and more lucrative, the latter is uncertain and, even at the end of World War II, remuneration was frugal. But research has other rewards. Sutherland was influenced by the great biochemist Carl Cori, a professor of biochemistry at Washington University, to enter research. Cori has enriched science by the guidance he gave to young, aspirant scientists. The biochemist Arthur Kornberg also did his apprenticeship in research in Cori's laboratory.

Sutherland began studying the effect of hormones on a system that was dear to Cori's heart. He and his wife, the late Gertie Cori, had shown that there is an enzyme which liberates glucose from glycogen. The first step in this liberation is phosphorylation, or the addition of a phosphate to glucose, forming glucose 1-phosphate. This reaction, as we said earlier, was known to be stimulated by adrenalin in the whole organism. The Coris had tried to study the effect of various hormones on this enzyme *in vitro*, but they failed.

Sutherland began a patient study of the effect of the hormone not on the intact liver in the whole animal but in very fine slices of liver through which the hormone could penetrate. He was able to observe the effect of the hormone in this semi–*in vitro* system. The amount of phosphorylated glucose increased after the administration of adrenalin. Sutherland next showed that the enzyme which achieves this chemical reaction, the addition of phosphate to glucose, is an intermediary carrier of the phosphate group. It itself has to be combined with phosphate before it becomes active.

At this stage Sutherland switched and took a bold step. He decided to study the phosphorylation of the enzyme itself, not in intact cells but in cells which he gently disintegrated. He was successful and, therefore, he persisted, moving on to the next level of fractionation of his disintegrated cells. His early success was obtained with a mixture which contained partly soluble and partly particulate portions of the cellular debris. He tried to clarify this sytem by eliminating the particulate cellular debris by centrifugation. At this point the hormone activity could not be detected with the soluble material alone. This was only temporarily discouraging because it gave a clue to the possible site of action of the hormone. When the centrifuged cellular debris was recombined with the clear solution, the hormonal activity was regained.

Sutherland thus discovered that the particulate mixture is the target of adrenalin activity. At this point he had an extraordinary stroke of good fortune. We may add to Napoleon's dictum: Not only are good generals lucky; good scientists are often the favorites of that fickle lady. Sutherland added adrenalin to the particulate debris and then made an extract of this mixture. Now he added this extract to the soluble fraction, and indeed he could show increased phosphorylation of the protein. He was luckier still. The extract was boiled and was still effective. This was a great advantage because boiling would coagulate and remove a lot of other inert materials present in such a complex mixture.

The next step followed the path of classical biochemistry. The heat stable extract was concentrated, fractionated by a variety of chemical methods; at each step its potency in activating phosphorylation *in vitro* was tested as a guide to the concentration. Inactive fractions were eliminated, active fractions concentrated. Eventually Sutherland and his associates were rewarded. They isolated a solid material which had the potency to activate phosphorylation in soluble solutions *in vitro*.

Chemical analysis revealed that the material contains adenine, the sugar ribose, and one phosphate group. Thus in composition it appeared to be an adenosine monophosphate, a well-known component of tissues, since it is the precursor from which ATP is made. But it did not behave chemically like any known adenosine monophosphate.

In order to identify his baffling product, Sutherland sought the aid of Dr. Leon Heppel of the National Institutes of Health. Dr. Heppel is one of our foremost experts on the chemistry of these compounds. Moreover, he is very generous with his expertise. Sutherland asked Heppel for some enzymes which are known to cleave phosphate groups, hoping thus to try to identify the product that remains. Heppel responded generously but he left Sutherland's letter on his desk. Some time later he was clearing his desk and read Sutherland's letter. Very close to it was another letter from a colleague, David Lipkin, who described to Heppel a novel reaction of ATP he discovered. When he allowed ATP to be in the presence of strong solutions of barium hydroxide, phosphate was precipitated out, but also an unusual product which contained adenine, ribose, and one phosphate emerged. Lipkin drew the conclusion that his product was adenosine monophosphate but an unusual one; it had a cyclic structure. Heppel concluded that the two findings described in the adjacent letters were the same substance, *cyclic* adenosine monophosphate (cyclic AMP), and he conveyed this information to Sutherland.

With the availability of abundant cyclic AMP from the chemical preparation, progress became very rapid. Sutherland determined that in the membrane component of the material that he sedimented by centrifugation there is an enzyme which can convert ATP into cyclic AMP. In turn, cyclic AMP stimulates the enzyme phosphorylase which is on theother side of the membrane within the liver cells.

From these clean-cut experiments the molecular mechanism of a particular hormone activity became apparent. Adrenalin awash in the bloodstream reaches the outside membrane of the liver cell. There it activates an enzyme, to which the name adenylcyclase has been given, to convert ATP into a cyclic AMP. The cyclic AMP enters the cell, stimulates the enzyme phosphorylase, and converts glycogen into glucose 1-phosphate, thus starting the first reaction in a whole series which can then provide energy in soluble form back into the bloodstream. Cyclic AMP acts as a kind of second messenger; the hormone adrenalin is, of course, the first. This is the system of relaying messages across the barrier of the cell membrane in this instance. This was but one molecular mechanism of hormone action, and physiologists catalog a host of hormones.

With the availability of cyclic AMP, a cheap product which soon became commercially available, a number of other investigators began to try to see whether cyclic AMP has any effect upon the enzymatic or hormonal systems of their immediate interest. Soon communications from all over the world on the effect of cyclic AMP began to pour into the journals, until at last count 40 different hormonal effects had been reported.

In most instances, cyclic AMP stimulates the activity of some enzyme. In others, the activity is more complex and the mechanism is unknown. For example, *Dictyostelium discoideum,* the slime mold whose differentiation we discussed earlier, is under the effect of cyclic AMP. When the individual slime mold amebas are well nourished they secrete into their external medium cyclic AMP, but at the same time they also secrete an enzyme which removes the phosphate group from cyclic AMP. When nutrients run out, the amebas are no longer capable of synthesizing this extracellular enzyme and, therefore, cyclic AMP accumulates in the external medium. This seems to be the signal for all the amebas to begin to gather together and to go through a variety of steps in differentiation which produce spores that may be deposited at some distance from the stalk, and thus the cycle may start again in a nutritionally richer area.

The use of this second messenger, cyclic AMP, in such divergent organisms as the slime mold and mammals is one more example of the frugality of Evolution's invention. Evolution is the supreme pragmatist: cyclic AMP is an effective messenger for the slime mold; it is preserved for the same role in mammal, or *vice versa.* Since we do not know the evolutionary history of the slime mold, its stimulant for aggregation may have been acquired long after the mammals started their evolutionary climb.

The mechanism of the accumulation and decomposition of cyclic AMP reveals why the effect of cyclic AMP in a cell is only temporary. This is essential, of course, if it is to be an effective second messenger, because if a cell's metabolism is to be kept under control the original signal cannot persist permanently. Indeed, we find in mammalian cells which contain cyclic AMP an enzyme which destroys cyclic AMP slowly after the message is received. These enzymes split the cyclic phosphate bond into the precursor molecule, adenosine, which contains no phosphates. (It is the same enzyme which the slime mold excretes to destroy cyclic AMP.)

The versatility of cyclic AMP was documented even further by Ira Pastan, a young man at the National Institutes of Health. He found that cyclic AMP can stimulate not only enzyme activity, in other words, the activity of preformed enzymes, but it can also enhance the synthesis of enzymes. Thus cyclic AMP is a potent agent in enhancing the production of the inducible enzyme which splits lactose, the much-studied enzyme we have discussed earlier. In this instance, cyclic AMP acts by increasing the production of messenger RNA for the synthesis of the enzyme.

Pastan is one of the vigorous investigators studying the effect of cyclic AMP on cancer cells as well. These studies are at a very preliminary stage, but it does seem to appear that cancer cells are low in cyclic AMP. Whether this is a universal attribute of all cancer cells remain to be seen. Moreover, we also do not know if the lowering of the cyclic AMP in some cancer cells is below a threshold value for activity. The level in normal cells may be in excess, and the lowering observed in cancer cells may be merely a reflection of their rapid growth.

There are also reports from Pastan's and other laboratories that cyclic AMP, when added in large amounts, can give cancer cells grown in tissue culture an appearance of normalcy. The significance, if any, of these demonstrations remains to be seen because, upon the removal of the excess cyclic AMP, the cells revert to their cancerous state. Therefore the cyclic AMP has only a superficial effect. It does not reach the determining mechanism which casts the cell into its abnormal condition.

Not all hormones use cyclic AMP as a second messenger. The sex hormones, for example, usually do not; their mechanism of action is entirely different. This finding emerged from the studies of a skilled and persistent biochemist at the University of Chicago Cancer Research Laboratories. The motivation for these studies came from the work of Dr. Charles Huggins of the same laboratories, who was the first to show the interrelationship between hormones and some tumor growth. Dr. Elwood Jensen decided, against the taboos of physiologists, to study the molecular mechanism of the action of a sex hormone. In order to do this, he synthesized some of the sex hormones with very high radioactivity incorporated in them, and then he began a patient study of what happens when the hormone is injected into an experimental animal.

A very complex mechanism emerged from Dr. Jensen's studies. In the organs which are the target for the hormone, a protein in the cell's peripheral membrane picks up the hormone and binds it to itself. This protein then carries the hormone across the cytoplasm. It is a ferry of a sort. When this cortege reaches its goal, the nucleus of the cell, the hormone is handed over to still another protein which is awaiting it at the inner border of the nucleus. Under the influence of the hormone, it is known that this target protein in the nucleus undergoes a structural change. In turn, a whole series of new molecular activities are triggered so that new RNAs, both messenger and transfer, are produced, presumably to begin synthesis of proteins specific for the organ controlled by that hormone.

Thus, as indicated earlier, if an immature rat whose uterus is undeveloped receives an injection of the appropriate hormone, estradiol, within 24 hours the primitive uterus will more than double in size and will take on the tissue characteristics of the mature organ. The exact molecular sequence which achieves this is not yet decoded. We know that brand new messenger as well as transfer

RNAs are produced. Therefore both the transcription and the translation processes must be altered to mobilize and express information present in the nuclei.

With the stimulus of hormones we can mobilize information from unusual tissues. Thus, for example, we can inject into a rooster huge doses of female sex hormones. The unfortunate creature begins to pour out of its liver a protein which is normally found only in the yolk of hen's eggs but is produced in the hen's liver. The poor rooster is totally confused at the sexual and molecular levels. The population of its transfer RNAs changes and, as we said earlier, different transfer RNAs can have different functions in the translation mechanism.

Hormones have a variety of structures. Some, such as insulin, the human growth hormone, and a few others, are proteins; some are smaller molecules containing but 6 to 8 amino acids; others are derived from simple amino acids; and a group of them belong to a chemical family called sterols. Just as varied as their structure are the many functions of hormones. Because of the complex structure of many hormones and of the complexity of the reactions by which they exert their magic influence, much of the molecular biology of hormonal action is still unknown.

A very good example of this complexity is the human growth hormone, which determines the height to which a child will grow. Absence of it for any reason, or its deficiency, will result in dwarfism in a child. The human growth hormone is one of the few hormones which is specific to the human species and is very complex in structure; therefore, until recently, replacement therapy for the dwarfed child was impossible. But, as a result of some good biochemistry and the cooperation of pathologists, remedy of this particular kind of dwarfism can be routine.

The human growth hormone is a protein containing some 190 amino acids. The sequence of these amino acids, as we said earlier, is unique and specific for humans. The growth hormone is produced along with other hormones in the pituitary, a gland located at the base of the brain. To add to the complication, it is known that the growth hormone is released from the pituitary gland by still another hormone called the growth hormone-releasing hormone. A variety of stimuli are known to release the growth hormone. Among them is low blood sugar; therefore children with a tendency toward dwarfism have been treated by the injection of insulin. Still another impetus for its release is sleep. This, of course, lends credence to the old wives' tale that children grow during their sleep.

The stimulation of growth hormone production by sleep is well documented. The amount of circulating hormone can be determined by its immunological properties in blood taken during waking and sleeping hours. In turn, this stimulus by sleep revealed the mechanism of the impact of emotional state on the growth of a child. One of our distinguished pediatricians, Dr. Henry Silver, had observed stunted growth in some emotionally deprived children. He associated this with the disturbed sleep of such children and, in turn, with a negative influence of disturbed sleep on the stimulation of growth hormone production. In this instance, we have a clear connection between the psychological state and the physical development of the child. This is a rare revelation of the molecular mechanism of the effect of "the psyche on the soma."

Absence of growth hormone or its partial deficiency may be due to two entirely different causes. In some cases it is the result of a benign tumor close to the pituitary gland. Such tumors may be removed surgically, and in such cases the production of the growth hormone sometimes resumes. In other cases the deficiency is due to a faulty development of the embryo. This probably has a genetic origin because 15 percent of children deficient in growth hormone also have a similarly affected sibling. In such cases there was no remedy until about 12 years ago when a campaign among pathologists was started in England to send human pituitary glands removed during autopsy to the Department of Biochemistry at Cambridge University where a competent biochemist would process the tissue and extract the hormone.

Pathologists in the United Kingdom send annually about 60,000 human pituitary glands for processing. After careful chemical extraction and purification there is enough material to treat 500 or so children with the hormone. Fortunately, since dwarfism is rare this seems to be adequate in the United Kingdom. Unfortunately, there is no such organized program in our country, nor is there any prospect of the production of synthetic material on a practical scale in the near future. The sequence of the amino acids in the hormone has been determined, and its synthesis is technically feasible but, since the need and market are small, production costs would be prohibitive.

After injection of the hormone, bone and cartilage begin to grow, through a mechanism whose molecular basis is totally obscure.

The effect of the hormone is nothing short of phenomenal. During two years of therapy a boy whose growth had been arrested will become more than a head taller, and this growth will continue until by the end of six years the child will achieve normal height and normal masculine appearance. Because of the complexity of the interactions of hormones, untreated male dwarfs tend to be undermasculinized as well. No hormone is an island unto itself.

We have discerned some of the regulatory mechanisms of living organisms, but many of us feel that there is still a great deal hidden from us. The paucity of knowledge and the challenges we face will become even clearer when we study not isolated systems of enzymes and hormones but the integration of all of them into the development of the individual, a topic which I shall present in the next chapter for the reader, who, I hope, is still with me.

The miracle is not a specific life;
the miracle is any life.

ANONYMOUS

Chapter six

Routine Miracles

An almost universal impulse of informed women upon the delivery
of their infants is to examine toes and fingers for reassurance that
the baby is normal. In 997 cases out of a thousand the mothers
can sigh with relief: Another miracle has run its nine-month course.
I use the term miracle advisedly because the current status of our
knowledge of the wondrous molecular events between fertilized
egg and the first breath of a new human is so limited we can
but contemplate the process with awe and admiration and applaud
the genius of Evolution.

We can split the atom, we can safely visit the moon, but of
the mechanism of differentiation we know next to nothing, and
of our cerebral processes—how I manage to line up a sequence
of words, and how the reader receives them so that we understand

123

each other, I hope, completely—of this we know nothing. Let us put our knowledge of differentiation in perspective with the splitting of the atom: We are probably at the stage where the English philosopher Dalton defined the atom in 1805. As far as our knowledge of mental functions goes, we are in the pre-Copernican era of cosmology. We know we are here, we presume we can think, and that is the sum total of our knowledge.*

We know enough about peripheral mechanisms of embryonic development to make it tantalizing and to yearn to explore the hidden inner secrets. We all start our life's journey as a fertilized human egg which is only 0.14 millimeter in diameter and is just barely visible to the naked eye. Such a fertilized egg is diploid. It contains the complement of chromosomes from each parent's sex cells. Impelled by unknown molecular signals, this cell begins to convert some of the material that it contains into the total genetic apparatus needed by a cell. The rest is a finely programmed sequence of events which can best be studied not in a dividing embryonic cell but in cells in tissue culture.

First comes a growth-1 period, G-1, which lasts about 4 hours. During this period there is extensive protein synthesis, probably in preparation for the synthesis of DNA, which occurs only during the period which has been designated the S or synthesis period. This is when DNA is replicated so that there will be enough for division into the two daughter cells. The S phase usually lasts about 7 hours. It is followed by a phase called G-2, a period of packaging of the DNA into about 1/10,000 of its length when synthesized. This is a tremendous feat of packaging because a DNA molecule, which when fully extended may be a meter long, is wrapped faultlessly into short, thick rods. The next period is cell

---

*To be sure, some people, usually physicists without experience in biology, and therefore without appreciation of the uncanny inventiveness of Nature, will come forward to proclaim that the human brain is merely a compact computer. The great biophysicist, the late Leo Szilard, once heard a presentation by some young man on the machinery of the brain. Szilard fidgeted but listened to the jargon of "computerese": input, output, loops, and bits; at the end Szilard lifted his bulk and said to the young man: "Maybe *your* brain works that way, but not mine!"

division, or the M phase, for mitosis; it takes about 1 hour. At the end of cell division in the embryo we have two identical daughter cells surrounded by an envelope. Should this envelope break for any reason and the two daughter cells separate, two identical individuals, identical twins, may be born.

In about 3 days after the first division there may be a hollow sphere of 32 cells. They are still surrounded by a membrane, the zona pellucida, and thus the young embryo is still not in communication with the blood supply of the mother. The cells are small in size. Some time during the 6th day after fertilization the free-floating globe of cells sheds its membrane and touches and adheres to the surface of the uterus. With uncanny foresight it chooses an area of the uterus which has capillaries beneath it. The nature of the molecular divining rod which detects the capillaries, even though they are hidden by a layer of other types of cells, is totally unknown.

Let us leave this early human embryo to its privacy now and look at the embryonic development of organisms which can be experimentally manipulated. Much of our knowledge of the development of an embryo comes from studies of nonmammalian species. Here the fertilized egg contains enough nutrients to support the developing embryo to the hatching stage. For the student of biology this is, of course, an enormous boon: We can observe the exposed developing embryo and can manipulate it mechanically and chemically. The former, mechanical manipulation, was performed with remarkable success by embryologists at the turn of the century.

A fertilized egg divides by logarithmic progression into 2, 4, 8, 16, 32, 64, etc., and forms a fluid-filled sphere, the blastocyst. It achieves a spherical shape by cleaving alternately in direction at right angles to the previous cleavage. Were it to continue growing this way, and at this rate, the human embryo in 9 months would be a monstrous ballon larger than the earth. This catastrophe is avoided by the first miracle of differentiation.

At a certain stage—when the number of cells reaches several hundred, called the blastula—some of the cells which appear to be identical, but are not, begin to flutter and also to divide faster. They could move either of two ways, outward, forming a fingerlike projection from the sphere, or inward, looking like a soft balloon, one side of which is pushed in. Evolution has eliminated any would-be organism that formed the fingerlike excrescence because under those conditions the cells would remain as they were in the hollow sphere, a single layer of cells interacting with each other in two dimensions only. As a result of the invagination, a third dimension is added and, as we shall see, this new environment has profound effects on development.

At what stage are the dividing cells rendered individualized by differentiation? Charles Weissmann, the genius who predicted reduction of chromosomes during meiosis, thought that each cell is unique even at the two-cell stage, each carrying, as we would say today, half of the information for the creation of the eventual individual. Weissmann's "mosaic hypothesis" even found confirmation. The embryologist Roux took a two-cell embryo of the frog, destroyed one cell with a hot needle, and lo! a monster half tadpole developed. This was one of the first examples of experimental embryology. More experiments designed with greater ingenuity and executed with finesse were to follow.

Another inquisitive scientist, Driesch, took a different approach to determine whether the dividing embryonic cells carry only part of the mosaic or are complete. He gently shook a two-cell stage of the sea urchin until it separated into two distinct cells. From these arose two normal free-swimming larvae. Driesch was able to do this with as late a stage of embryonic development as 16 cells, when, shaken apart vigorously, 16 normal larvae emerged from the 16-cell stage of the embryo. This experiment, of course, proved the equality in information of all of these cells. One does not know how Weissman took this news, but geniuses too can err, or get lost in the maze of the complexity of biology.

What went wrong with Roux's experiment? His strategy and conclusion were faulty, the monster arose because its development was impeded by the scarred, cindered debris which had remained attached to the live cell.

When do the cells in the hollow sphere lose their totipotency? When do they become individualized? The answer to this question came from the laboratory of a German scientist, Hans Spemann, who, during the turmoil of World War I, continued, with patience, his questing experiments. From visual observation of the day-to-day development of the early blastula, it became obvious to him that cells were not randomly distributed; they had an invariable topography. It became apparent, for example, that certain cells give rise to a neural plate, or crest, from which the central nervous system eventually develops.

Spemann and his students developed some exquisitely delicate operative techniques. They would excise certain cells from one part of the blastula and graft them onto the distal side of the sphere. Thus, when they took a group of cells which were presumably going to develop into the neural plate and grafted them onto the belly side of another blastula of the same age, the graft developed not in accordance with its origin but rather in an identical way to the cells of its new surroundings. These were very revealing experiments. Apparently the environment of the cells has a controlling effect upon the eventual fate of the transplanted cells.

One of Spemann's students, Hilde Mangold, decided to perform similar experiments in later stages of development of the embryo. She transplanted a very small piece of the top of one embryo into the belly side of another one of the same age and carefully followed the result of her delicate operation. She observed a striking phenomenon. A second neural plate was formed on the belly of the recipient. It developed into a neural groove and neural tubes so that on both the back side and the belly side of this creature neural cords were developing. Dr. Mangold chose her experimental embryos well. The donor organisms were rather pale in color and the recipient organism's cells were much darker. Therefore she was able to observe that not only the few cells that she transplanted to the bottom of the recipient continued to form the neural plate, but also that many of the cells of the recipient host differentiated into neural plate.

Two very important conclusions could be drawn from the skillfully executed experiments of Dr. Mangold. In the first place, one could conclude that after a certain time, or after a certain number of cell divisions, the cells are committed to a specific function in forming the individual. After decades of study we feel today that this commitment, once made, is irrevocable. The possible exceptions are cancer cells, but as we shall discuss in the next chapter some experts believe that cancer cells do not reverse their commitment; they had not become differentiated in the first place.

The second conclusion was more important because it was not a static statement but rather a stimulating prediction from which further experiments could lead to deeper insights into embryonic development.

The cells on the belly of the recipient embryo were captured by the few transplanted cells and were diverted from their usual, normal goal. Development into neural cells was foisted upon them. Spemann thought that some substance, an inducer or "organizer," must diffuse out of the committed neural cells and divert the neighboring recipient cells away from their original commitment. The hunt was on for the organizer. First it was determined that either the organizer had a miraculously broad spectrum of activity or there were several organizers. These conclusions came from new experiments designed after the pattern of Dr. Mangold's on developing embryos. Some organizer was able to induce the production of head structures; still others were able to induce the production of tail structures.

Under Spemann's influence, biochemists were induced to try to determine the nature of the organizer but the history of this research is lugubrious. Dozens of claims were made for the isolation of the "true organizer," none of which could stand repetition in other laboratories. In defense of the many scientists who spent years on this problem, it should be stated that it was an extremely difficult one. The amounts of starting materials available are very small because the organizers have such potent effects that they are present in but minute amounts, often in but a few cells.

We know now that very early in embryonic development, sometimes in the 8-cell stage, only some of the cells have specific enzymes and enzyme products. This information has emerged from the development of a new field of ultrasensitive chemistry, cytochemistry as it is called. The Danish biochemist, Lindenstrom-Lang, and the American, Oliver Lowry, were the leaders in perfecting these ingenious methods, and with them we can see enzymatic differentiation in cells which cannot be distinguished by any other means in the very early embryo. Therefore the synthesis of some of these enzymes must occur very early.

We can also detect very early potent chemicals such as serotonin, which in the adult is known to control blood pressure and ryhthmic movement of the cells in the lining of the intestine. What the contribution of serotonin is to the development of the embryo we do not know. However, it must have some function because chemicals which interfere with serotonin function in the adult interfere with the development of the early embryo.

One substance that might be considered an organizer is within reach of being identified. It is erythropoietin, the substance which stimulates the synthesis of red cells. Erythropoietin is potent in extremely low concentrations, and in this respect it qualifies both as a hormone and as an organizer. Studies of erythropoietin can be approached because two requirements absolutely essential for such research are fulfilled. Tremendous amounts of starting material, tons of beef blood, are available and there is a test for its activity. When erythropoietin is injected either into immature bone marrow or into fetal liver, the production of red cells is stimulated.

The difficulty in the search for organizers was highlighted by the work of the embryologist Vogt. He developed an ingenious method for tracing the path of certain cells from the early embryo to their eventual location in the developed organism. He marked individual cells with a dye that did not kill the cell but adhered to it throughout its journey. He accomplished this by pressing small lumps of agar impregnated with various dyes against a few cells on the outer surface of the blastula. With this method, Vogt was able to determine the ultimate location of various cells. He actually worked out what is known as the "fate map" of an embryo by locating the origin of various differentiated tissues or organs.

Vogt's achievement was extraordinary in the early 1920s, and we have not improved on it. Today we can label cells with radioactive compounds in their DNA and remove these cells from a labeled donor and graft them onto an unlabeled recipient and follow the path of the cells via the radioactivity. But the results are essentially the same as those obtained by Vogt's simple methods of dyeing the cells. From these and other studies, the extraordinary complexity of embryogenesis emerged.

Cells committed to the formation of a certain organ, for example the heart, very often travel huge distances compared to their size to get to their committed destination to form the appropriate organ. Moreover, from other studies which we shall relate later, we can estimate the paucity of cells which form a given tissue or organ. For example, we can calculate that all of the pigment cells in the hair of the mouse originate from perhaps as few as 34 cells of the early embryo. In view of this we should be patient with the slow progress in searching out the secrets of embryonic development. The organizer for the formation of hair itself would have to be sought in about 170 cells in any one embryo. Should we ever succeed in isolating an effective inducer of hair formation, we could put the toupee manufacturers out of business.

The peripatetic migration of cells during embryonic development discovered by Vogt has been visually confirmed by recent investigations which use some refined cinematography. The outstanding investigator in this field is a Swedish embryologist, Tryggve Gustafson. He takes advantage of the transparency of the sea urchin embryo which, in the blastula stage, consists of a hollow sphere one cell layer thick. Gustafson focuses his microscope on a sea urchin blastula and then attaches to the microscope a movie camera equipped with a time-lapse device. When the developed film is run rapidly, the series of photographs taken at intervals reveal a pulsating motion of some of the cells.

In the blastula the cells are attached to a membrane which envelops them from the outside. As a result of the vibrations, the cells tend to round up, presumably reducing the surface of attachment to the membrane, freeing the cells from the glutinous envelope. Soon a new type of movement appears. The cells form long, thin protrusions or pseudopods. These improvised legs move around randomly almost as if they were exploring the inner cavity. In time, the projections make contact with the inner wall of the blastula and, if the contact is appropriate and propitious, the pseudopods contract, pulling the cell along with them. The method of recognition for the appropriate site where the migrating cell will settle is unknown.

But another extraordinary complexity in the mechanism of differentiation is clearly revealed. Not only must migrating cells be equipped with some recognition system, but also in terms of embryonic size a few recipient cells, maybe a mile away, must have the lock to the migrating cell's key. We do know, however, that there are certain proteins, so-called aggregating factors, which can coalesce cells of homogeneous structure and function. Thus we can take cells in the early embryo which can be recognized as a primitive heart and, if these cells are gently disintegrated, their reaggregation can be studied. If cells of another type are mixed in with these heart cells, they are excluded and slowly the cells of the heart find each other, aggregate via some glutinous protein which is the aggregating factor, and such a clump of cells when it reaches a sufficient size *in vitro* begins to pulsate like a heart.

During the infolding of the blastula three different types of tissues develop. They have been given the names ectoderm, endoderm, and mesoderm. We know from fate maps and from other evidence what kinds of organs each of these archetype cells gives rise to. The ectoderm is the origin of skin, hair, and nails. The mesoderm gives rise to muscles, our circulatory system, bones, kidneys, and the reproductive system. The endoderm gives rise to the lining of the digestive system and the bladder.

The embryonic developement of the eye is an extraordinary con-

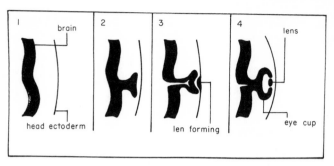

Figure 6.1. Schematic Diagram of an Organizer at Work

catenation of interaction of tissues and molecular forces. For this, two different archetype cells must cooperate. The eye cup originates from the ectoderm and it pushes forward to reach toward the surface of the head. The cup eventually develops into the retina and also into the various folds of the eye. Meanwhile, the part of the epidermis in front of the eye cup thickens and through a series of differentiating steps becomes the lens. This is an amazing achievement, involving as it does the joining of two different types of tissues on an exquisitely timed schedule and the structuring of tissues to appropriate shapes and sizes with extraordinary precision.

The construction of the lens is under the control of the optic cup. As this cup touches the epidermis, some kind of stimulus is transferred which induces the epidermal covering cells to develop into the lens (Figure 6.1). The control of lens development by the optic cup has been demonstrated by the skillful surgery of the embryologist. If the eye cup is removed the lens does not develop at all, or one can remove the epidermis which normally would form the lens and replace it with a piece of epidermis taken from another part of the body, from the head or even from the belly, and this foreign, transplanted tissue, once it is in contact with the optic cup, develops into a lens.

If a thin layer of cellophane is inserted between the optic cup and the epidermis, no lens develops. However, if a porous membrane is inserted between the eye cup and the epidermis, a lens does develop. This would seem to indicate that the inducer, or the organizer for the lens, is some material which can diffuse through the agar and, therefore, it is a relatively small molecule. The exact nature of this material has been probed by tracer methodology such as introducing radioactive amino acids to the eye cup and following the radioactivity to the eye lens. However, such experiments are suspect because the amino acid could diffuse naturally without being organized into a protein.

These then are the problems, both technical and theoretical, which the experimental embryologist faces. It is obvious there is not one organizer, as Spemann had visualized; there must be several. It is also obvious that the organizer oozing out of the eye cup must be highly specific to command the exquisite differentiation of tissue into the highly specialized lens, which must be perfect both in shape and in transparency. To achieve the latter even the nuclei of the cells must be induced to disappear.

What happens to organizers after the appropriate organs are developed is not known. In some lower organisms they may persist because they can regenerate lost limbs and other organs. Thus, if the eye lens of the adult newt is removed, the newt will form a new one by inducing another nearby tissue to go through the complex series of differentiating steps which produce the lens. Imagine the value of this capability were we to possess it. A lens with cataracts which destroy its transparency could be removed, as we do now, but we could regrow it! Why does not an organism whose organizer for the eye lens is retained into adulthood have trouble because of the continuing synthesis of several eye lenses? We do not know. We have to resort to an unproved, improvised explanation: While it is intact, the lens exudes some inhibitor of lens development.

Very few regenerative powers are possessed by mammals; skin, hair, nail, and red cells are obviously regenerating constantly. The healing of wounds must be considered a regenerative function. But the most unusual regeneration in mammals is that of the liver. Three-quarters of the liver of an experimental animal can be surgically removed and within a couple of days the organ is restored to its original size and shape. On the other hand, if part of a kidney is removed it does not regenerate to its former shape. If one of the kidneys is completely removed, the remaining one will grow to twice its former size. This is the reason donation of a kidney for transplantation can be done with impunity. The molecular signal which stimulates the remaining kidney to grow to assume the new load of filtration all by itself is unknown.

Since the timing of the assembly of different tissues into close proximity is so precise and the control of tissues over each other is achieved with such minute quantities of potent agents, it is not at all surprising that if foreign agents intrude they can wreak havoc with the normal development of the embryo. An excellent example of this was provided a few years ago by the indiscriminate use of the drug thalidomide.

Ours is an era of drugs. Not only of drugs used for escape from reality but of drugs which are doctor-prescribed for escape from pain, anxiety, or insomnia. The number and variety of organic chemicals we swallow for a variety of ends has been climbing steadily in recent years. The reasons for this are many. Drug companies throughout the world are putting huge efforts into searches for new molecular combinations to alter the functioning of various organs: the brain—make it calmer; the kidneys—make them excrete better; the heart—dilate its constricted vessels. Once some success is achieved some drug companies encourage physicians to use their products through blatant advertising or personal calls by "detail" men. The physician, under pressure from his patients for new miracle drugs, unfortunately, often yields. The number and variety of pills consumed by some affluent members of almost all societies is frightening. To be sure, some of these are beneficial and essential, but many are superfluous agents of pampering.

Thalidomide, a drug with tranquilizing effects, was developed by a German drug house and it rapidly gained popularity. Fortunately, our own Food and Drug Administration examines new drugs quite thoroughly before licensing their use. Such caution has been rewarded. Americans were not permitted to switch to this new tranquilizer. (Hopping from drug to drug is a rather typical neurotic pastime.)

After the introduction of thalidomide into the drug market, infants with monstrous deformities began to be born in Europe as well as in Japan, Peru, Kenya, and Lebanon (an indication of the universality of the drug habit). Some infants had mere stubs for limbs; others were even more mutilated. The origin of these unfortunates was traced to the consumption of the drug by the pregnant mothers. The disaster, in addition to serving as a warning against "pill-hopping" during pregnancy, is also serving another useful purpose. Thalidomide has become a useful tool for the study of teratology, the development of monsters. Thalidomide produces the same ghastly malformations in monkeys as in humans. Therefore the effect of the drug, both in dosage and in time of administration during the pregnancy, can be studied. (Curiously, thalidomide is not teratogenic in rats.)

Perhaps the best demonstration of the delicate balance of the internal and external environments required for normal development of the embryo is that oxygen, the life giver, can be toxic. About 40 years ago the incidence of blindness among prematurely born infants was observed with increasing frequency. The origin of the malformation, to which the name retrolental fibroplasia was given, was obscure until a bright physiologist, Dr. Leona Zacharias, got onto the trail of the etiology of the disease.

First of all, she correlated the occurrence of this type of blindness with the birth of the premature infant in large communities. "Preemies," as the nurses like to call them, in rural areas escaped this blight. From this correlation she moved to the next one that the affected were born in large hospitals with excellent facilities for nurturing preemies. The continuation of her detective work led to the study of the oxygen concentrations in incubators in which premature babies were kept. As an aid to the premature human who is struggling for life, oxygen concentrations in such incubators were kept more than twice that in the normal atmosphere. And that turned out to be the source of the blindness in such infants!

In the uterus, the oxygen supply of a developing fetus is controlled at a steady level. Under these conditions, certain blood vessels which are needed for the development of the eye but are not needed later are reabsorbed by the developing fetus. When the fetus is expelled prematurely from this salubrious environment and is given a high concentration of oxygen to breathe, the reabsorption of these vessels is arrested and, therefore, transparency is either partially or completely lost. Oxygen concentration in incubators for premature babies is, of course, now very carefully controlled, and this particular malformation has returned to its usual low frequency.

Whenever ideas and novel experimental probings are exhausted in a field, almost invariably an investigator with an original turn of mind will enter it and with ingenuity and persistence perfect new approaches and thus open new frontiers. It is obvious that embryology studied with the tools of chemistry has bogged down. Those tools as they exist today have yielded the maximum information to be pried from the mystery of the developing embryo. But a new system, which any number of us would have predicted could not work, has been introduced into embryology. Dr. Beatrice Mintz, of the Fox Chase Cancer Institute of Philadelphia, can create embryos and newborn mice not from the usual endowment of two parents but from four of them.

Dr. Beatrice Mintz is a beautiful, dark-haired woman who, should one encounter her away from her laboratory, one would think is a petite flamenco dancer in civvies. However, a very brief conversation would reveal that she is a scientist of extraordinary ability who also has a beautiful facility for expressing and describing her complicated experiments and profound conclusions. About 10 years ago she undertook bringing together two early embryos of mice. She chose mice with clear differences in genotype which could be readily seen in their phenotype. The embryos she chose were from pure black mice and pure white. She removed two very early embryos from the two different strains of pregnant mice. She obtained best results in her experiments with embryos in the

8-cell stage and, therefore, this is the one which we shall describe.

At this stage of development of the embryo it is still covered, as we said earlier, by a thick membrane, the zona pellucida. Dr. Mintz decided to try to peel off the zona pellucida not mechanically but with a protein-splitting enzyme. There is a very good commercial product, pronase, which is a protein-splitting enzyme that is isolated from bacteria. She very carefully immersed the embryos in the 8-cell stage in a solution of pronase under carefully controlled conditions of enzyme concentration, acidity, and duration. She observed that the zona pellucida fell away but the 8 cells remained adhered to each other. She washed off the enzyme from the embryo with great finesse so that the 8 cells were neither damaged nor dispersed.

She now placed two such preparations of embryos, one from a pure black strain and one from a pure white strain, in close contact in a tiny dish which was specially coated so that the eggs would not adhere and merely pushed the two embryos together with a glass rod. She observed that after about 10 minutes the two embryos miraculously fused and 24 hours later formed a composite sphere. A female virgin white mouse was induced to become "pseudo-pregnant" by mating it with a male on which a vasectomy had been performed, and into the uterus of this foster mother Dr. Mintz implanted surgically the embryo of complex ancestry which she had manipulated. She then provided the best care possible for the female mouse which was tricked into this complicated pregnancy. Dr. Mintz was extraordinarily fortunate because in many cases she could obtain live mice from normal delivery at the end of gestation.

The progeny turned out to be neither black nor white nor a fusion of the two colors but striped! In Figure 6.2 the strategy of the experiment is shown and in Figure 6.3 a product of such an experiment. This mouse had not two parents but four; not four grandparents but eight, plus, of course, an "incubator" mother.

The first time in history that such a mouse with such parentage was born, and survived to old age, was in 1965. Since then hundreds of normal animals of similar parentage have been produced by

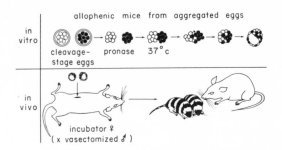

Figure 6.2. Schematic Diagram of the Production of Mice
with Eight Grandparents

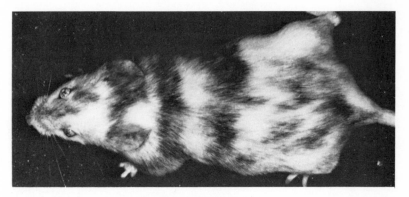

Figure 6.3. A Mintz-Made Mouse

Dr. Mintz's methods, and the number of normal offspring produced in turn by normal matings runs into the tens of thousands. Dr. Mintz has coined the term "allophenic" for these mice; in other words, the individuals have a simultaneous orderly manifestation of two different cellular phenotypes, each with a known distinctive genetic basis, such as black and white.*

Allophenic mice have proved themselves a remarkable tool for the study of the development of individuals. Dr. Mintz's work and the conclusions she has drawn deserve a whole book in them-

The offspring of the mating of these mice are normal and follow the usual Mendelian pattern of inheritance. In other words, their sex cells are normally derived from one or the other kind of germ cells.

selves, but we shall list here only a few of the highlights. The simplest explanation for the colored bands is that each of these bands descended from a single pigment cell, or a single cell which is committed to pigment formation. The band of homogeneous color then arises from what Dr. Mintz calls a clone. This is a term from bacterial genetics for a group of cells all originating from the same cell and having identical genetic endowment. She can count 17 clones on each side of the mouse and, therefore, we may conclude that the primordial cells which are destined to give rise to pigment cells are 34 in number and that they remain stable through uncounted cell generations until the individual is born. The direction of the pigment cells during growth is well defined. They seem to proliferate chiefly laterally rather than logitudinally, hence the long, narrow, horizontal stripes.

This is the first time we have had such an insight into the process of embryonic development. We can get other information with even more subtle markers or the phenotypes that are expressed. Thus we can study the mode of formation of other tissues and organs. Markers here have to be more sophisticated than the easily seen pigment. Dr. Mintz used two different lines of mice in which some specific enzyme differs in physical characteristics so they can be separated in the laboratory in an electric field. When such mice with different enzyme patterns are mated normally, the offspring has a new type of enzyme which is a hybrid. This hybrid enzyme is present in both the muscle and the liver of such a normal offspring. However, if allophenic mice are manipulated from two such different backgrounds, their liver does not contain the hybrid enzyme; only their muscles do. From this we may draw the conclusion that liver and muscle are formed differently during embryogenesis.

We know from studies of muscle cells that these elongated cells have more than one nucleus. The question is whether these multi-nucleated cells are formed by the division of cells and their lack of separation or whether they are formed by the fusion of neighboring cells with single nuclei. Dr. Mintz's intricate but well-designed experiments gave the answer. It is the latter. Only in this way could hybrid enzymes be formed within the cytoplasm of a muscle cell by random fusion of cells with different genetic information.

The number of questions that can be posed for exploration with allophenic mice is legion. For example, we can prepare allophenic mice from strains which show a high frequency of mammary cancer and strains with a low frequency of these cancers and study the factors which may determine the emergence of the cancer in the offspring. Indeed, the interaction of cells bearing almost any variable condition can be explored.

In the tradition of great scientists, Dr. Mintz is generous in teaching her ingeniously refined techniques to young colleagues who travel to her laboratory from all over the world, and these young people in turn are asking still newer questions. In the next few years their answers will be surfacing in the scientific journals of the world.

The reader may be surprised that a biochemist has said so little about our knowledge of the molecular biology in the developing embryo. This is certainly not due to modesty, it is rather, to paraphrase Winston Churchhill, that we have a great deal to be modest about. The fact is that our knowledge of the molecular forces in embryonic development is miniscule. We know that RNA, DNA, and new proteins are produced. We also know that at certain phases of the development of the embryo the production of brand new proteins is turned on. Indeed, we believe that differentiation is achieved by the production of new proteins. But the mechanism of cueing in and out of the synthesis of RNA and specific proteins is totally obscure.

We use all the tools in our armamentarium of biochemistry to pry out answers but the answers are often incomplete, and very unsatisfying. We know, for example, that, if the early sea urchin embryo is immersed in a drug called actinomycin D, which is known to inhibit the transcription of DNA into messenger RNA, the embryo, nevertheless, continues to divide and can reach the free-swimming larval state. The conclusion drawn from these experiments is that in the sea urchin no new messenger RNA is needed in the fertilized egg. The messenger RNAs are apparently there and only their translation waits to be switched on. This is a very unsatisfying conclusion because we know that actinomycin D does not shut off all the transcription of DNA. A certain amount of it might escape inhibition by the drug. Indeed, it has been shown very recently by Dr. Claude Villee of Harvard that there is some RNA synthesis as early as in the 2-cell stage of the dividing fertilized sea urchin egg. However, it is not messenger RNA but new transfer RNA which is newly modified by appropriate methylating enzymes as early as in the 2-cell stage.

This may be a concomitance unrelated to further development but I would, of course, doubt this. I take recourse in the dictum of Sir Henry Dale. When contemporary physiologists scoffed at his claim that there is a chemical messenger, acetyl choline, between nerve endings and muscle which actually transmits the message of the nerve, Sir Henry said, "God did not put acetyl choline in the synapse merely to fool the physiologists."

The transfer RNA must be formed and shaped so early to be a new agent of translation of previously laid down messenger RNA. But how the molecule to be translated from the thousands there is chosen is a total mystery.

This fragment of knowledge about the new transfer RNA in the 2-cell stage is tantalizing and frustrating because our current chemical tools can take us no further. Thus we know enough to know how little we know.

Some of the questions posed by Aristotle are essentially still unanswered. He wrote in the *Generation of Animals*:

How, then, are the other parts formed? Either they are all formed simultaneously—heart, lung, liver, eye, and the rest of them—or successively, as we read in the poems ascribed to Orpheus, where he says that the process by which an animal is formed resembles the plaiting of a net. As for simultaneous formation of the parts, our senses tell us plainly that this does not happen: some of the parts are clearly to be seen present in the embryo while others are not . . . . Since one part, then, comes earlier and another later, is it the case of A fashions B and

that it is there on account of B which is next to it, or is it rather
the case that B is formed after A? . . .

In the early stages the parts are all traced out in outline; later on
they get their various colours and softness and hardnesses, for all the
world as if a painter were at work on them, the painter being Nature.
Painters, as we know, first of all sketch in the figure of the animal
in outline, and after that go on to apply the colors.

Are we to be pessimistic about ever penetrating the mystery
of the "routine miracles"? I do not think so, but it is better to
admit our current ignorance than to hide it under a net of discon-
nected, hazy glimpses of parts of the majestic performance. It is
a wise man who knows what he does not know.

And loathsome canker lives in sweetest bud.
SHAKESPEARE

Chapter seven

Growth Without Shape: Cancer

What is cancer? I will spare the reader recounting, as so many historians of medicine do, the antiquity of cancer. The opening of Pandora's box occurred so long ago and surviving visible human remnants are so young, it would be astounding if the inhabitants of the valley of the Nile of some 5000 years ago had *not* been plagued by our contemporary ailments. Five thousand years may be a long time to the student of recorded history, yet it is but a fleeting moment in the long history of man's evolution and of his coexistence with the parasites—friendly and malignant—that he harbors. There is no reason to suppose, nor any evidence at hand, that our very recent ancestors were any more sturdy in structure or more resistant to infection than we are today.

The parasites of various sizes from worms to viruses which inhabit

143

our bodies today were probably with us not for 5000 years but for 100,000 and more and, considering the furious ingenuity with which man is working on making himself an extinct biological curiosity, the chances are they will be here to mourn the passing of the last of our species.

Nor will I try to define cancer in medical terms. Often cancer specialists cannot agree among themselves on a succinct definition. I will try to define cancer as a cell biologist and a molecular biologist see it.

The methods of culturing mammalian cells outside of the body which we have described in earlier chapters enable us to look at cancer not as a mass of tissue which will eventually strangle the individual but as single cells. How do cancer cells behave compared to normal cells when they are isolated and placed in appropriate media by themselves?

It has been known for a long time that profligate motility is an abnormality, characteristic of the cancer cell. In a normal individual, usually only the red and white blood cells are mobile. Other than that, tissue cells do not move. We have known that the cancer cells defy this restriction; they can move and reestablish themselves in sites distant from the original source and continue to divide and multiply and produce new large masses, very often of the original tissue type. Thus it is not unusual to find cancerous thyroid tissue in a part of the body distant from the thyroid gland.

On the other hand, normal cells such as fibroblasts (elongated cells from skin) when isolated and grown on a glass surface migrate actively. The membranes of such cells are constantly undulating. The wavelike motion at its crest can be as high as 5 microns. However, when the ruffling membrane comes in contact with the membranes of a similar fibroblast cell, the wavelike motion ceases. The two cells remain in close proximity but all movement halts. Thus the glass surface soon becomes covered by the cells dividing and migrating in directions away from the original site of seeding, and we observe a monolayer lawn of cells covering the glass surface.

The term "contact inhibition" has been given to this behavior of normal cells which prevents overcrowding on a plate. Now let us place a tumor cell into tissue culture medium on a glass plate and observe its behavior.

First of all, under a microscope the tumor cell exhibits an entirely different shape. It is more rounded and its membranes do not undulate in the same direction. Instead, leglike projections move out from the periphery of the tumor cell and the cell moves in the direction of the projection. Thus the appearance of the cancer cell confirms the name which had been almost intuitively given to cancer, that is, the "crab." When a cancer cell is seen in stereovision the beholder shrinks away from it in horror (Figure 7.1).

As these cells keep multiplying and begin to cover the plate, an extraordinary property of cancer cells becomes apparent. They are not inhibited on contact. They pile up on top of each other in almost random fashion. The cancer cells have lost their capability of contact inhibition. There is a good correlation between the ease with which the cancer spreads in the original host and the extent of its loss of contact inhibition. We think that this attribute of the surface of the cancer cell—for that is where signals of proximity are received—may be the explanation for the ability of cancer cells to invade normal tissues. In Figure 7.2a and 7.2b normal mouse cells and cells converted to malignancy with consequent loss of contact inhibition are shown.

That the surfaces of cancer cells are different from those of normal cells has recently been shown chemically as well. There are certain plant products, so-called "lectins," which adhere to the surface of mammalian cells, and it is found that more of these lectins can adhere to a cancer cell than to normal cells, with consequent clumping. There are more sites of receptivity for these products in a cancer cell than in normal cells (Figure 7.3).

There is still another line of evidence which proves that the surface of the cancer cell is different from that of normal cells. The cancer cell has immunological differences. There is either some protein or polysugar or the combination of the two on the surface

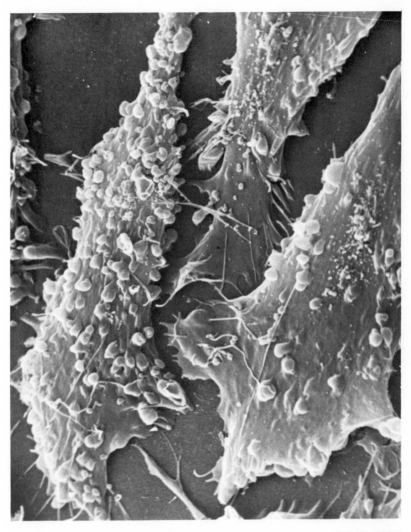

KEITH R. PORTER

Figure 7.1. Cancerous Cells

of a cancer cell which, when injected into a rabbit, produces
antibodies that are unique to the cancer cell. These antigens, whatever they may be, are called the "surface antigens" of the cancer

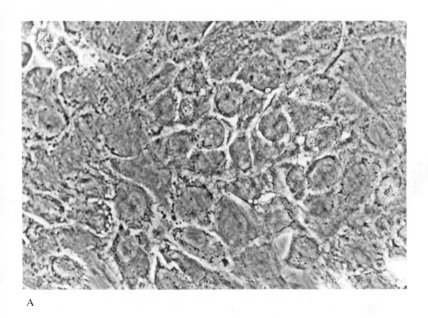

A

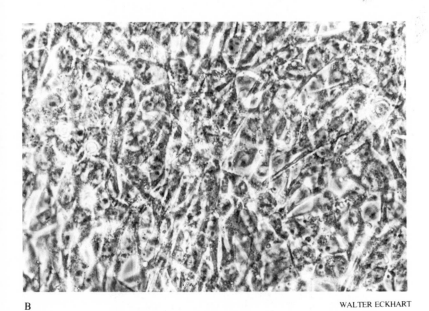

B WALTER ECKHART

Figure 7.2. A Normal Mouse Cells B Cancerous Mouse Cells

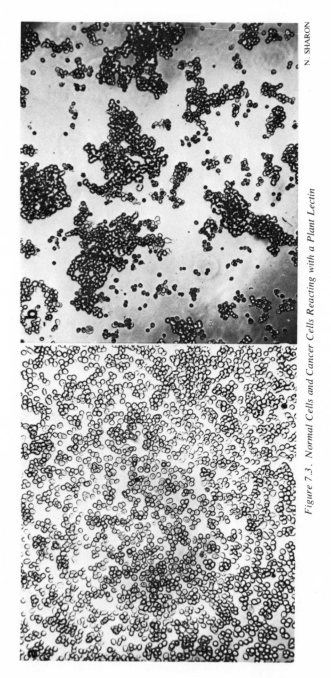

Figure 7.3. Normal Cells and Cancer Cells Reacting with a Plant Lectin

cell. This is but one kind of antigenic difference between cancer and normal cells. There are others. Indeed, in some instances, especially those cancers provoked by certain chemicals, an antigen unique to that particular cancer is observed.

Studies of the immunological properties of cancer cells are relatively recent, on the order of some 14 years, but they are accumulating a great deal of evidence for the immunological foreignness of cancer cells. In turn, these studies have revealed why cancers may not be contagious. In the early days of cancer research, investigators connected the circulation of a subject with leukemia to the circulation of a volunteer normal individual, usually an inmate of a prison, and even though the leukemic cells were introduced into the circulation of the normal subject he did not come down with leukemia; his body rejected the abnormal cells.

In another study, aging patients who had some debilitating but noncancerous ailments were given transplants of human tumors and it was found that they rejected the transplanted tumors. These studies evoked, very properly, loud protests, for it was not clear that the aging indigents were aware that the procedures being followed were not done for their own welfare, for therapeutic reasons, but rather for experimental purposes. These studies, like almost all studies in our country, are supported by Federal funds; when they surfaced, the Federal government drew up a code of conduct for studies involving human patients: such studies cannot be performed today without the informed consent of the patient.

In turn, it is well known that patients with certain cancers such as Hodgkin's disease have a very limited immune response to other inciting agents such as the tubercle bacillus.

Thus there is an immunological protective mechanism in normal individuals which rejects or defeats cancerous growths. Indeed, there is a hypothesis first proposed in 1959 by a Dr. L. Thomas but made popular by Sir McFarlane Burnett, the Australian immunologist, that the immune system which animals have developed emerged not for the protection of the organism against invading infectious agents but to protect the organism against spontaneous cancer cells. According to this hypothesis, each of us may have cancer cells arising within our body with high frequency, but the system of immunological surveillance detects these potentially lethal cellular accidents and eliminates them.

This hypothesis must be considered but a stimulating speculation since it can be neither proved nor disproved. It merely speculates on evolutionary determinism of millions of years ago. However, cancer immunology is being intensely studied both for a possibility of cancer prevention and for diagnosis. For example, there is a certain circulating antigen in the bloodstream of subjects who have cancer of the colon. Unfortunately, however, the correlation between this antigen and the cancer is not absolute. Moreover, in at least one instance, that of a tumor of chickens which causes economic disasters, the tumor's growth can be prevented by immunization. But, as we shall see later, this immunization is not against the cancer itself but against the virus which produces the cancer in the chicken.

Are there any known differences between the molecules present in cancer cells and those in normal cells? A number of differences in enzyme activity are well known, but in most of them the difference is not *qualitative* or absolute; there are more or less enzymes with certain activities. Thus, for example, for a long time an observation and a hypothesis introduced by the German biochemist, Otto Warburg, held sway. He was the first to study the source of energy in a cancer cell in a highly refined apparatus he devised. He noted that the cancer cell derives most of its energy from the first stages in the metabolism of glucose, stages which require no oxygen. This pathway releases only part of the energy of glucose. Presumably, this mechanism operated in primitive life forms before the earth had its blanket of oxygen. (Our envelope of oxygen is presumed to have been acquired later from the activity of photosynthetic plants.)

Warburg postulated that the cancer cell is a primitive cell which returns to this primitive type of respiration without oxygen for its source of energy. He postulated that the conversion of a normal tissue into cancer is simply a switch due to such a trauma as contusion, which turns off the pathway of utilization requiring oxygen. Since Warburg was equally accomplished as a biochemist and a polemicist of giant proportions, this hypothesis, however naive, held sway for a long time even though it was blatantly simplistic. It neglected all of our knowledge of genetics and also the increasing evidence that viruses can be causal agents in cancer.

In the late 1930s a very exciting claim came from the laboratory of the German organic chemist, Kögl. He and one of his associates claimed that there is an absolute difference between an amino acid from a tumor and the same amino acid from normal tissue. Nineteen of the 20 amino acids which compose our proteins can exist in two forms. The difference between the two forms is very subtle and slight. It centers upon the orientation of one carbon atom, the one to which the amino group is attached. Spatial distribution of four different components around that carbon atom can be different, forming mirror images of each other.

Through a method of selection which is totally obscure to us, living organisms have evolved to use amino acids with the same pattern of distribution around the pivotal carbon atom. Kögl and his associate claimed that they isolated from tumors one amino acid, glutamic acid, which did not have the normal configuration. It belonged to the type not found in normal tissues. This was electrifying news since for the first time it implied a fundamental biochemical difference between normal and tumor tissue. Alas, this putative difference did not stand scrutiny. Other biochemists isolated the same amino acid from tumor tissue and could not confirm Kögl's sensational claim. The source of this blunder is totally obscure.

However, today we can state with confidence that there is a biochemical component which is altered in every cancer cell examined. In each of these cancer cells there are qualitative alterations in transfer RNA, the cardinal component of the protein-synthesizing machinery described in an earlier chapter. How these alterations in transfer RNA in cancer cells affect them is not clear at present.

When we discovered that the modifications of transfer RNA are species- and organ-specific, I suggested that they must serve a function for organ- or species-specific tasks, such as the regulation of the synthesis of proteins. I also suggested that these modifying enzymes, which introduce methyl groups, may be aberrant or abnormal in cancer. Some forty different cancers have been examined all over the world and crude enzymes isolated from tumor tissue in every case do function abnormally. In turn, if these enzymes which modify the transfer RNAs were abnormal I suggested that the products, the transfer RNAs themselves, might be different from those in normal cells. This hypothesis too has been tested in laboratories all over the world, and in every cancer cell tested so far there are transfer RNAs which are not present in a normal tissue.

These differences are *qualitative* biochemical differences between a cancer cell and a normal cell. It is not that there are *more* transfer RNAs but *different* transfer RNAs. Whether these transfer RNAs were woven differently or altered differently by the enzymes that we discussed is not known yet, but it will be soon. Whatever the outcome of these investigations, there is now no doubt that the protein-synthesizing machinery in every cancer cell has different components which are unique to it. How such unusual transfer RNAs function in cancer cells also requires extensive research, but, as we stated earlier, it is now known that in bacteria a very slight modification of transfer RNA can keep the production of as many as 9 different enzymes switched on permanently.

What does the molecular biologist know of the cause of cancer? Of the many reputed causes of cancer, chemical carcinogenesis has the most ancient history. It was observed by Sir Percival Pott in the eighteenth century that cancer of the scrotum seems to be an occupational hazard of the chimney sweeps in England. He very astutely correlated the heavy coat of tar on the bodies of these men with the cancer which some of them developed. Interest in the possibility that chemicals in our environment are the cause of cancer flared up sporadically thereafter. There was a flurry of excitement in the 1930s when it was shown by English workers that the purified components of tar which can produce cancer in animals have as a kernel of their structure an organic chemical configuration which is also present in the kernel of the structure of cholesterol, the lipid material which is present in every cell. It was simplistically concluded that some aberration of cholesterol metabolism causes cancer. But, unfortunately, from a considerable effort to track down this hypothesis nothing conclusive emerged.

Nevertheless, stimulated by these findings a great many workers, especially in England, studied a variety of chemical compounds as possible cancer-producing agents and by now they have found a great many; indeed, the list of chemicals which can produce cancer is frighteningly long. Some of these are man-made; some of them abound in nature. Thus a few years back whole flocks of turkeys in England were dying of cancer. The outbreak of the disease was correlated with the feeding of peanut meal imported from Africa. Some astute biochemical detection discovered that the carcinogenic component of the peanut meal was a compound, aflatoxin, which is produced by a mold that makes peanut meal its habitat.

Liberal consumption of peanut products, such as peanut oil, has been suggested as a possible cause of the high frequency of cancer of the stomach among Koreans. This conjecture, which must remain unconfirmed, highlights the difficulty of establishing unequivocal causality of cancer in humans. It is unthinkable to put 20 Korean children on a diet rich in moldy peanut products and 20 on diets free of the suspected carcinogen and observe them to adulthood.

It is unthinkable to perform such an experiment willfully, but on rare occasion such an unwitting experiment on humans can be studied, provided an imaginative and determined investigator chances upon the trail. Dr. Howard Ulfelder, Professor of Gynecology at Harvard, eminently fulfills these requirements. Dr. Ulfelder, who is an outstanding specialist in cancers of the female organs, diagnosed within a brief period cancers of the vagina in seven teen-aged girls. Such cancers in that age group are rare; a busy, successful gynecologist might easily have dismissed them as a quixotic negation of the laws of probability. But, fortunately, Dr. Ulfelder is not only busy and successful; he is also a scholar whose mind is as broad ranging in interest as it is precise. The mothers of two of the girls told Dr. Ulfelder that they had been given stilbestrol during their pregnancies with the afflicted girls.

Some 30 years ago a synthetic female hormone, diethylstilbestrol, was introduced into medical practice. It was convenient to administer orally, and it was recommended for a variety of female problems including miscarriage. A quick survey revealed that the mothers of all seven girls had taken diethylstilbestrol during their pregnancy! After Dr. Ulfelder announced this possible correlation, scores of teen-age girls with vaginal cancers were found to have been born from diethylstilbestrol-managed pregnancies. These correlations are frightening, for they highlight the complexity of the cancer-producing mechanism and the baffling tasks cancer researchers often face. The man-made hormone analog alters the female embryo in the uterus in some subtle manner; the effect remains unrecognized for fifteen or more years, only to burst forth as a ravaging malignancy.

Unfortunately, diethylstilbestrol is part of the environment of not only pregnant women but of all of us in our country. Chickens receive implants of diethylstilbestrol under their skin, and calves are fed the drug because it is an effective fattening agent. The remnant of the pellet is supposed to be discarded after the slaughter of the animal and, hopefully, the residual drug in the tissues is eliminated during cooking, but I for one would not bet on it. After considerable resistance from cattlemen, the feeding of diethylstilbestrol has been finally banned by the Food and Drug Administration.

The implication that viruses are the causal agents in cancer has a much shorter history. The French scientist, Augustus Borrel, suggested as early as 1903 that viruses may have a role in cancer. He had no evidence for this. It was strictly an intuitive observation after he noted that viruses can cause increased cell multiplication. Five years later two German scientists, Ellerman and Bang, reported that they could produce leukemia in fowl by an agent extracted from afflicted animals which was not filterable. But not much attention was paid to this because in those days the status of our knowledge of malignancy was such that it was not yet accepted that leukemia is a malignancy of the blood-forming apparatus.

In 1911, Peyton Rous, who worked at Rockefeller Institute, reported the results of his investigation on fowl sarcoma which did cause widespread interest, but mostly of a derogatory nature. Rous stated that he could take macerated tissue of a fowl sarcoma and pass it through one of the finest filters, which was known to hold back all infectious agents except viruses, and upon the injection of such a filtrate into healthy fowl he observed the emergence of the same kind of tumor as in the donor. Rous's claim was simply rejected by the then ruling circles of cancer pathologists.

It was not until 1932 that a virus was accepted as an etiologic agent of cancer when Richard Shope, also an investigator at Rockefeller Institute, showed with impeccable experimental evidence that the rabbit papilloma, a tumor of rabbits, can be transmitted into other rabbits via a scrupulously filtered extract of the tumor. By 1932, the technology of biological research had improved, but more importantly the minds of the new generation of investigators were more ready to receive ''new'' ideas even though the ideas were 20 years old.

A little later, in 1936, Bittner showed unequivocally that breast cancer in certain strains of mice is transmitted from mother to suckling mice by a filterable factor present in the milk of the mother. If the mice were wet-nursed by mice without the high frequency of cancer, the offspring escaped the disease. Bittner went further and was able to show via the electron microscope the presence of globular particles in the milk of the mice with the high incidence of cancer. Somewhat later, in 1951, Ludwig Gross reported that he could produce leukemia in a strain of mice via a filterable factor or virus present in the bloodstream of the afflicted.

Another event occurred in 1951 across the ocean, in France, which was to revolutionize our ideas about viruses and the hosts they invade and subjugate.

Louis Pasteur, the uncannily talented genius of biology, was also endowed with the practical turn of mind of his race. This was evident in the immediate application of his scientific findings to practical use, such as how to grow silkworms, or how to improve the beer and the wines of France. He organized an institute which bears his name. He visualized it as an enterprise which would carry on the researches he inaugurated, but he also thought of the ways to finance such a perpetual enterprise.

Part of the efforts of the institution were to be devoted to generating funds by the sale of biological products which were the fruit of his research efforts and those of his successors. As it turned out, the institution was only partially successful in generating enough funds to keep the scientific enterprise going. But the scientific leadership provided by members of the Pasteur Institute, especially in recent years, justified the hopes of the venerated founder. For example, at this institution the bacterial viruses, the bacteriophages, were discovered by the French-Canadian investigator, d'Herelle. Three decades of research achieved an understanding of the mechanism of action of these miniscule enemies of bacteria.

Phages consist of an outer coat made of proteins which extends into a tail or a needle. This bag contains either DNA or, in a few instances, RNA. The phage attaches itself to a receptive host bacterial cell and injects via its tail the whole nucleic acid content of the body of the virus. Once within the cell, by a complex variety of molecular mechanisms, the phage DNA or RNA captures the machinery of the host cells and subjugates it to synthesize not the substance of the host cell but rather the substance of the invading virus. After varied lengths of time the virus turns on the production of an enzyme which dissolves the bacterial cell wall, and out pour as many as 100 or more identical copies of the original virus.

All this we knew by the mid 1940s, but the origin of the first invading virus remained a subject of controversy. This controversy reached back 30 years to d'Herelle himself, who was the first one to suggest that not only do bacterial viruses invade bacteria but also some bacteria can give rise to such viruses spontaneously. D'Herelle was challenged and vilified by some of his contemporary colleagues. However, the concept that certain bacteria exist which are "lysogenic," that is, they are dissolved by some internal virus, was not abandoned.

Two prominent workers also at the Pasteur Institute who were pursuing this idea were Eugene and Elizabeth Wollman. Ideas, like wealth, are sometimes inherited. Unfortunately, their research and their lives were abruptly ended when the Nazis captured Paris and carried off the Wollmans to an extermination camp—because they were of the Jewish faith. It is invidious to single out two souls for mourning out of the millions who thus perished. But the loss of the Wollmans is particularly pertinent to our story for in their studies of latent viruses they were approaching, some of us think, the secret of cancer itself.

After the war Dr. André Lwoff, also of the Pasteur Institute and a close friend of the Wollmans, took up the problem of these self-destroying bacteria. First of all, he addressed himself to the moot question whether such viruses can indeed arise spontaneously. Up to then it could not be excluded that the virus kills a few cells in a large population of bacteria constantly. Thus a small number of bacteriophage could coexist with a large population of its prospective victims.* By then scientific technology had advanced to a stage where Lwoff could undertake an experiment to resolve this question unequivocally.

As stated earlier, there had been devised micro manipulators

---

*There is an obvious mathematical fallacy in this supposition. Bacteria reproduce by division into 2, let us say, every 20 minutes, but during the same period a phage can be replicated 100-fold or, to put it mathematically, the multiplication of bacteria is a function of 2^n and of phage it is of 100^n, where n is the number of generations. Obviously, if any free phage were let loose in a population of bacteria susceptible to it, the bacteria would soon be wiped out.

which through a system of reducing gears diminish the movement of the human arm so that minute forces can be imparted to equipment such that a single bacterium can be sucked up into a tiny tube. Lwoff selected a strain of microorganisms which were putatively lysogenic; some bacterial virus was always associated with them. He placed one of these organisms into a microdrop of their nutrient medium and observed it under a microscope for an hour, at which time it divided.

Lwoff seized one of the two daughter cells with his micro manipulator and transplanted it under the microscope into a fresh sterile droplet of medium which contained neither bacterium nor virus. He again observed the bacterium for an hour and witnessed the division of the parent cell into two. He continued this manipulation patiently for 19 generations, which was the end of his endurance: 19 hours. It was now evident that no phage could have multiplied within a bacterial cell in 19 generations for no cell was killed; every one was accounted for. Nor could any original phage be carried along in 19 transfers.

If we assume a tenfold dilution in each transfer, then there was a 10^{19}-fold dilution of the original solution at the start of this virtuoso experiment. In turn, it is obvious that 10^{19} phages would have had to be present initially; that way 1 phage might have persisted through all those dilutions. That is, of course, impossible inasmuch as we can detect a single phage and, moreover, 10^{19} phages are a visible lump. Lwoff carefully grew a large culture from the single cell of the 19th transfer and began a study of these organisms with the impeccable ancestry. He soon found that, as the organisms grew—hold your breath, Dear Reader, for we are about to witness spontaneous generation of a sort—bacteriophage appeared among them.

Lwoff's first hypothesis was that an abnormal metabolism of some of these bacteria might have given rise to the phage. However exciting this discovery was, it could not be approached as a real biological phenomenon because the frequency of the event was low; it could be calculated that it was on the order of about 10 per 100 cells. Lwoff persisted patiently and boldly. If 10 out of a 100 cells can be pushed on a destructive path, why cannot they all be induced to do the same? Through one of those intuitive conclusions, the process of which none of us can understand, Lwoff hit upon the expedient of exposing the lysogenic organisms to a small dose of ultraviolet irradiation. He described with charm what happened next, and, since this was one of the really important biological experiments of the century, the reader may like to be a witness to it.

Lwoff had a technical assistant, a lovely French girl, Evelyne Ritz. She so identified herself with her work that the bacterial culture and she became one. Fifteen minutes after the irradiation she said to Lwoff, who had collapsed in his chair "in sweat, despair, and hope," "Monsieur, I am growing normally." This was devastating news because if Evelyne and the bacteria were growing normally, the irradiation could not have had any effect on the bacteria. After another 15 minutes, Evelyne came in again and reported simply that she was still growing. Abysmal failure!

Finally an hour after the irradiation she came in and said quietly, "Monsieur, I am completely lysed." The bacteria disappeared! (I ought to explain the "growth" of Evelyne, or rather of the bacteria after the irradiation which was so devastating to Lwoff. The growth of bacteria can be determined electronically by measuring their turbidity. Evelyne's bacteria were not growing by cell division, but rather their size was increasing, producing increased turbidity, as the phage developed.)

Patient measurements showed that 99 percent of the bacteria lysed and there was a corresponding increase in the number of bacteriophages in the medium. Lwoff tried other agents, x-rays, and some chemical compounds which are known to be cancer producers. Invariably, these agents caused the activation of the latent virus within the bacterial cell, producing virulent virus which in turn could attack other bacterial cells.

These simple but brilliantly conceived experiments opened up a whole new frontier for the study of interrelationships between host cells and latent viruses. The field soon became clarified through the efforts of Lwoff and a few of his students. Viruses can be of two kinds, virulent, which upon infection destroy the whole cell, or they can be "temperate."

If the temperate viruses invade a cell, they may either kill it or they may become integrated into the genetic material of the host and be reproduced indefinitely along with the host through its numerous cell divisions and may give rise to virulent virus only through some accident of metabolism or exposure to some "inducing agent," all of which, as we said above, turned out to be carcinogens as well. There is, of course, tremendous survival value in the integration of the phage into the genes of bacteria. If phages could only invade a bacterial cell and kill it, bacteria and phages would soon become extinct. As a friend of mine, a distinguished virologist puts it, "It's a stupid parasite that kills its host."

That the temperate virus was actually part of the genetic material of the host was demonstrated in brilliant studies by François Jacob, whom the reader has met earlier, and by Elie Wollman, the son of the martyred scientists of the Pasteur Institute.

Lwoff understood clearly the implications of his findings to cancer. He would come into the laboratory, pick up a test tube culture of lysogenic organisms, pause as if in deep thought and would say, "These bacteria have cancer," and then he would walk away.

It would be pleasant to report that the seminal experiments and conclusions stirred the imagination of biological scientists and cancer researchers. Nothing like that happened. Studies of the lysogenic mechanism were carried on by but a few biochemists and geneticists alert to the significance of these findings. Lwoff, who is a stimulating lecturer, was not even permitted to visit our country for a couple of years during the McCarthy era.

Fortunately, Wendell M. Stanley, the man who first isolated a virus in relatively pure form, and who was a great popularizer of science, happened to visit the Pasteur Institute in the mid 1950s. He understood at once the relevance of latent viruses in bacteria to possible cancer formation by latent viruses of animals and he lectured extensively on these ideas, of course, always giving Lwoff full credit for his seminal influence. However, sources of ideas can vanish in the antiquity of a decade, and old ideas may be newly discovered and refurbished with new names.

In the late 1960s the "oncogene" hypothesis made its appearance. This is nothing more than an elaboration of the ideas of Lwoff and Wendell Stanley, embellished with some of the ideas of the Jacob-Monod hypothesis such as regulatory genes and structural genes and repressor systems. Simply stated, the oncogene hypothesis postulates that every one of our cells contains an integrated virus which was acquired by infection millions of years ago and has been carried along by vertical transmission from parent to child for eons. When certain external agents, such as a chemical carcinogen or even some other virus, encounters this ancestral seed of destruction, it can become activated, producing cancer by giving rise to a variety of proteins which convert the cell to malignancy.

An enterprising young English investigator, Robin Weiss, has been in the forefront in "inducing" virus from apparently virus-free chicken tissue by the methods that Lwoff had used. However, the obvious objection to this procedure is the high probability that the domestic chicken acquired these viruses from its close association with man and other domestic animals. Weiss pursued his studies on the wild jungle fowl of Malaysia, the putative progenitor of our contemporary chickens. He found that these free-living creatures also carry latent viruses which are amenable to "induction."

One very obvious barrier to credence in this hypothesis was the discovery of a number of viruses which produce tumors in animals which contain not DNA as their genetic material but RNA. How could an RNA tumor virus persist in the genetic material of an animal which is, of course, composed of DNA? This paradox has been resolved by a brilliant discovery by two young American cancer specialists. Dr. Howard Temin, of the University of Wisconsin, postulated that enzymes may exist which convert the information of an RNA virus into DNA and, in turn, this DNA may become integrated into the genetic material of the mammalian host. His evidence for this hypothesis was scanty when he searched for such enzymes in mammalian cells.

A few years later both Temin and Dr. David Baltimore of MIT hit upon a more abundant source of such an enzyme. They found that, when RNA-containing tumor viruses are grown in large amounts and the purified virus itself is examined, an enzyme which converts RNA into DNA is indeed present within the virus particles themselves. This enzyme, which was nicknamed "reverse transcriptase," enables the RNA virus to be duplicated into DNA; in other words, it reverses the normal flow of information which goes from DNA to RNA.

There was a flurry of excitement after the discovery of reverse transcriptase. There were jubilant shouts from young investigators that the " 'dogma' of Crick has cracked." Actually, the discovery of reverse transcriptase was but a partial negation of the principles brilliantly synthesized by Crick. Information is stored in DNA, it flows to RNA and from there to protein. The information from DNA to RNA goes via base complementarity. From RNA to protein a translation step is needed because those two systems have entirely different structures. That the information flow between DNA and RNA is reversible with a special enzyme does not negate the original concept of information transfer by base complementarity. Should we find an enzyme which translates protein sequence into nucleic acid sequence, that would be an enzyme of miraculous versatility.

Therefore the mechanism for investing the information of an RNA tumor virus in the DNA of the host certainly exists.

That viruses can produce tumors in animals has been proven beyond the shadow of a doubt. Indeed, there are scores of different viruses, made of either RNA or DNA, which are known to produce tumors in experimental animals. The evidence in these cases is unequivocal because it is relatively easy to carry out one of the famous tenets of Koch. Koch was the great German bacteriologist, one of the pioneers who demonstrated the bacterial origin of some diseases. One of his requirements for such a demonstration was the ability to isolate an agent of infection, which when injected into a receptive animal produced a disease. Koch's postulate is relatively easy to carry out in animals. It is easy to inject 100 highly inbred mice with a certain dose of a pure virus and then follow the mice throughout their life span.

Whether viruses cause cancers in man is not so easily demonstrated experimentally. In this case, we must draw conclusions inferentially. Man is not too different from other primates in his susceptibility to viruses. Variations of response, however, do exist. The best example of this is what is now known as the B virus. In 1920, Alexander, King of Greece, was bitten by a pet monkey. The wound became infected, fever developed, paralysis of the legs followed, and King Alexander died. The cause of his death was simply ascribed to septicemia, or blood poisoning. We do not know how many similar accidents occurred during the next decade—after all, "monkey bites man," unless he is a king, is not news.

On October 22, 1932, a Dr. William Brebner, a young Canadian physician working at New York University Medical College, was also bitten by an apparently healthy experimental rhesus monkey. Dr. Brebner was not unduly alarmed. As was the custom in those days, he dabbed the wound with iodine and went on with his work. The monkey died during a subsequent operation and the carcass was disposed of. Three days after the bite Dr. Brebner's hand showed symptoms of infection. There were little blisters around the bite; the wound and surrounding area were inflamed. In those days nothing much could be done to intervene in the course the infection was taking: there was no penicillin, no sulfa drug. A few days later, Dr. Brebner was hospitalized for he had a fever and developed increasing paralysis of the lower extremities. The paralysis advanced upward and on November 8, 17 days after the bite, Dr. Brebner died.

Fortunately, at this time a first-rate investigator of human viruses, Dr. Albert Sabin, was on hand. He undertook an attempt to isolate some infectious agent. Various organs of Dr. Brebner were emulsified and were injected into monkey, dog, mice, guinea pig, and rabbit. The only animal which became afflicted by the injection of extracts of the cadaver was the rabbit. The infection could be passed from rabbit to rabbit as many as 15 times. In each case the disease ran its remorseless course: paralysis to death.

The agent of the infection was shown to be a filterable virus but apparently not one of the known ones. In tracking down the identity of the virus, Dr. Sabin eliminated polio because, even though the original source of the infection was a monkey, injection into monkeys was innocuous. Rabies was similarly eliminated because the dog, the classic host of that disease, was immune.

Microscopic examination of some of the infected tissues of the rabbit revealed telltale granules in the nuclei of infected cells. Only three or four viruses were known to produce such granules. Among these is herpes simplex, the virus which is the agent of the common cold sore. Herpes simplex causes a relatively mild infection. About 70 percent of the human population acquire this virus by infection very early in life, and subject and virus live in symbiosis for a lifetime.

Certain stresses, such as exposure to large doses of ultraviolet irradiation, or a severe cold, or sometimes dietary idiosyncracies such as eating shrimp or lobster, will cause an eruption of the sore, usually on the lip, but sometimes in other parts of the body too. One can remove a small amount of fluid from a running sore, inject it into hen's eggs; the virus reproduces and thus it can be accumulated and studied easily. Its various properties can be studied and the virus can be "fingerprinted" and, as has been done recently, photographed under the electron microscope (Figure 7.4).

The herpes simplex virus is an icosahedron; it is a spherical structure composed of 20 contiguous triangles. The reader may recall that Buckminster Fuller invented the geodesic structure built of contiguous triangles which can be built to any size without internal support. Herpes simplex has priority on that invention by millions of years.

The virus that killed Dr. Brebner is very similar. It has exactly the same appearance as the herpes simplex of man, and it has

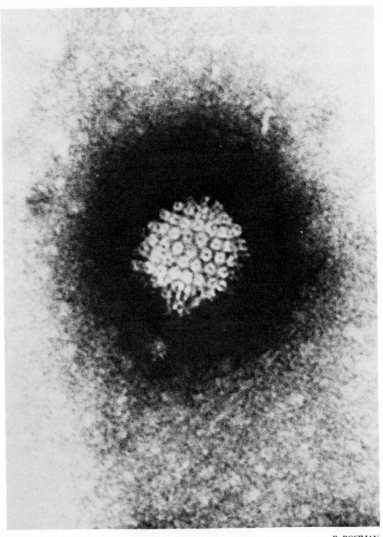

B. ROIZMAN

Figure 7.4. A Herpes Virus

all the other fingerprint properties common to it. But through eons of evolution these two viruses, the herpes simplex of man and the B virus of monkey—B virus was the name given to this virus after its tragic victim, Dr. Brebner—have become attenuated so that they do not kill the host. This again confirms the validity of the truism, "It is a stupid parasite which kills its host."

However, when we meddle with the pattern traced by Nature and reunite man and monkey in the laboratory of the virologist, a virulence of tragic intensity ensues. Since the death of Dr. Brebner well over a score of similar accidents have been recorded. The monkey bite almost invariably ends in the death of the victim. Interestingly, however, man's herpes simplex is innocuous when injected into the monkey. "Man bites monkey" is not news.

Still another surprise was awaiting us about the varied forms in which herpes virus lurks. Dr. Denis Burkitt is a British physician who, after service in Africa during World War II, returned to Africa essentially as a medical missionary. There he observed the prevalence of an unusual tumor among children. The tumor manifests itself by a monstrous growth, usually along the jaw of the afflicted. The growth is so rapid the patients often die within a few weeks after the onset of the disease. Dr. Burkitt observed that this particular disease occurs in a well-defined region of Africa where the climate is both hot and humid. The suspicion was forced upon investigators that Burkitt's lymphoma, as the disease came to be called, may be of infectious origin spread by some insect which thrives in those regions.

A young patient was flown to our National Cancer Institute in Bethesda, and there Dr. Sarah Stewart took some tissue from the patient during an operation and injected some of the material into guinea pigs whose thymus had been removed to suppress their immune defense. Dr. Stewart observed foci of disease akin to virus-produced tumors. Two English scientists, Epstein and Barr, went further. They cultured in tissue culture Burkitt's lymphoma, and they were able to show the proliferation of a virus which grew on the cells. The name EB virus, for the initials of the two scientists, has been given to the virus. Fingerprinting and electron microscopic examination revealed to our great surprise that it is another herpes type of virus very similar to herpes simplex, differing only in the identity of a couple of the dozen or so proteins present in the virus. This is the clearest case of the association of a virus with a human tumor, because EB virus, when injected into hamsters, invariably produces tumors.

Still another devastating disease in which a herpes type of virus appears to be the causal agent is one that afflicts chickens. Occasionally whole flocks of chickens are destroyed by an infection which may kill the animals immediately or may produce tumors among them. A virus was isolated and shown to be the etiologic agent of Marek's disease, a virus very similar to the herpes simplex of man. As we mentioned earlier, in this instance the disease can be prevented by protecting young chicks with an injection of the same virus grown in turkeys. This, of course, is another example of the individuality of virulence; a herpes of the turkey is innocuous to the chicken. But the two viruses are sufficiently similar that the antibody elicited by one protects against the other.

Still another tumor in which a herpes type of virus is implicated is the Lucké virus tumor of frogs. Under natural conditions this virus produces tumors only in the kidney of the leopard frog.

Another disease which is almost certainly caused by EB virus was identified by chance. Fortunately, the alert investigators recognized the meaning of their chance finding. Drs. W. and G. Henle are a husband and wife team of virologists who were studying the occurrence of antibodies to the EB virus in humans. For controls they used the blood of one of their technicians whose antibody level against EB virus was essentially zero. Suddenly, the antibody level circulating in her blood rose. This was alarming because how she became infected by EB virus was not known. Was she a victim of a laboratory accident?

The Henles correlated the sudden rise in antibody in the young lady with her having come down with infectious mononucleosis. This is a relatively mild, usually transient disease of the white cells of adolescents and young adults. It has been called the "kissing disease" because, presumably, infection is spread via saliva. This is not entirely accurate. Sometimes infectious mononucleosis may spread through a whole college dormitory from infection probably from unsterilized utensils.

Two different studies have confirmed the EB virus as the causal agent in infectious mononucleosis. At Yale University a sample of the sera of entering students is stored in a serum bank. The sera of students later stricken with infectious mononucleosis were tested for the presence of antibodies to EB virus. They were all positive. The sera of the same students in the serum bank taken before the onset of infectious mononucleosis were also now tested. These were negative. Therefore the antibodies to the EB virus were acquired either during or subsequent to the bout with infectious mononucleosis.

There was still another confirmation of the causal link between EB virus and infectious mononucleosis. Some young people have antibodies to EB virus which they may have acquired during a mild bout of the disease of which they had not been aware. Students with such positive anti-EB virus sera were observed for 4 to 8 years, and none of them came down with infectious mononucleosis during that period. On the other hand, 10 to 15 percent of the students who were originally negative in antibody to this virus contracted disease and their serum became positive with antibody.

These findings highlight the complexity of interactions of viruses with host cells. A very similar group of viruses produces a spectrum of disease among man, monkey, frog, turkey, and chicken. The differences among the viruses as seen by biochemical fingerprinting are slight, yet differences in virulence are enormous. We have not even an inkling about the sources of these life- or death-conferring biological mechanisms. We know a little bit about attenuation of virulence in the bacteriophage. In these much simpler systems we could demonstrate that the infected bacterial cell has defense mechanisms, enzymes which modify and degrade the DNA of the invading virus. Mammalian cells and mammalian viruses are orders of magnitude more complex. Our efforts to master their intricacies must be also orders of magnitude greater.

As we stated in the beginning of this chapter, the unanswered questions about cancer in general and viruses in cancer specifically are legion. The only thing common to all types of cancer is that there is some loss of control, and therefore more of the cell's genetic information is being expressed than in a normal cell. This can result in macabre growths, the teratomas, which may have hair growing in the center of the liver, or nails in the testes. Whether this loss of control occurred after the cancer cell had been differentiated into a normal cell and then became deranged and "de-differentiated" or whether a primitive cell, in an early phase of cell division, became cancerous and did not go through the process of differentiation, is a moot question which scientists are debating with words and attempting to answer by experiments.

The cytologist recognizes over a hundred different cell types in humans, and the regulatory mechanisms in each may vary slightly—some of these cells are under hormonal control; some, such as skin cells, are under the influence of environmental effects. Derangements of these many types of cells may occur in a hundred different ways. The pathologist recognizes about a hundred different types of cancers.

We must not assume that loss of regulation came about the same way in every one of these types of cancer cells. Loss of regulation may have occurred at the transcription level of the genetic material into information-bearing RNA, or it may have occurred at the translation level where that information is converted into proteins whose appearance or disappearance is basically the means of expressing differentiation. Therefore we may conclude that cancer is not a hideous disease such as syphilis or leprosy. It is a disease resulting from the loss of regulation.

Must we wait until the molecular biologist slowly decodes each of these regulatory mechanisms for a cure or palliation of cancer? Since there are over 600,000 new cancer cases a year in the United States alone, it would be tragic if the only hope we could offer the victims is patience. Fortunately, we can offer more than that. Of the 600,000 cases we mentioned, about a third can be cured. Surgery and irridiations have been used effectively for a long time.* However, these measures which essentially extirpate the cancer cells are useless in diffused diseases such as leukemia.

Fortunately, in 1955 the National Cancer Institute in Washington launched a program of therapy for cancer by chemicals. At first it was a modest program but it was very rapidly expanded by the largesse of Congress (indeed, too rapidly), and therefore in the early stages it was exposed to considerable criticism. Some of these attitudes persist even today among the ignorant who are not aware of the triumphs of this chemotherapy program.

The magnitude of the problem can be appreciated when we consider that there are a hundred different cancers of man and over a thousand of experimental animals, and the number of chemical compounds that the organic chemist has concocted and those found in Nature run into well over several million. A vast program of

---

*However, for success in therapy it is essential that the radiation be handled by knowing hands directed by expert minds. An instructive example is the x-ray therapy devised by the great biophysicist, the late Leo Szilard, whom we encountered earlier. Leo Szilard was one of the immigrants from Hungary to our country before World War II who enriched our intellectual life and also rallied to the defense of our country; he was a member of the team responsible for the atom bomb. Drs. Wigner, Teller, von Neuman, and Szilard were all Hungarian intellectuals. It was Szilard who convinced Einstein to inform President Franklin D. Roosevelt, through a friend, of the feasibility of an atom bomb.

Szilard made many contributions to physics, but after the war he switched his interest and became a first-rate biological scientist. Szilard's originality extended to every phase of his life. He was a character. When physicians diagnosed that he had a cancer of the bladder and wanted to operate, Szilard said: "There will be no operation until I consult *the* outstanding expert on tumors of the bladder." Szilard closeted himself for days in the library of a famous cancer institute. Later he assembled the physicians and they asked, "Have you consulted with the great expert?" Szilard said he had and there would be no operation. There would be x-ray therapy. The physicians asked who this greatest of experts was and Szilard said, with Hungarian modesty: "I am." He then directed the x-ray therapy with technical skill. His symptoms disappeared and he directed that upon his death an autopsy be performed upon him. He continued with his usual life of the intellect and of the body. Szilard was grossly stout. He lived in pastry shops. One of his favorite haunts was an Austrian coffee and pastry shop in New York where he would sit for hours reading, drinking coffee, and devouring huge quantities of pastry. After his death the autopsy was duly performed and the bladder tumor had indeed been cured. Szilard died of a coronary occlusion. The multitude of Sacher tortas and dobos tortas had exacted their toll.

screening, much of it under contract to industry, was set up. Likely compounds were tested for their toxicity in normal cells, and those that passed certain criteria were tested against tumor cells in tissue culture. From there the screening was carried on against tumors in animals and finally, slowly, the compounds were tested against tumors in humans.

These drugs were supplied to cancer specialists all over the country with appropriate directions for their use. There is a vast network of cooperating experimental cancer specialists organized under the very able leadership of the Director of Cancer Treatment, Dr. Gordon Zubrod; hence experiences can be rapidly exchanged and new information distributed throughout the country. This industry and this organization are yielding rich rewards, for today we can state that ten different disseminated cancers are amenable to chemotherapy, and the patient can have his normal life expectancy returned to him.

Response to therapy is a function of the speed of growth of the cancer. The cancer cell is most amenable to attack during its division, and rapidly growing cancers are thus the ones which are most prone to attack. The most spectacular response to chemotherapy is given in Burkitt's lymphoma. As we said earlier, this is a tragic disease of extraordinarily rapid growth. A single dose of a drug known as cytoxan* completely cures the disease in 60 percent of the patients without any evidence of later recurrence. If anyone questions the cost of the chemotherapeutic research of the past decade, let him look at Figure 7.5. Should one of my grandchildren come down with this dread disease and I could confidently take the child to an expert center of chemotherapy, I would say that the cost was more than justified.

*"Cytoxan" is the trade name of a compound which is essentially a modified nitrogen mustard. Nitrogen mustard is a derivative of the mustard gas the Germans developed to incapacitate and kill in the trenches of World War I. When administered in small doses, nitrogen mustard has been effective in controlling some cancers. This was the impetus for a small German drug house, "Asta," to prepare variants of it, including cytoxan. Its remarkable effectiveness in managing Burkitt's lymphoma was discovered by two American experts in cancer treatment, Drs. Herbert F. Oettgen and Joseph H. Burchenal of the Sloan-Kettering Institute of New York, who went to Nairobi to collaborate with Dr. Burkitt.

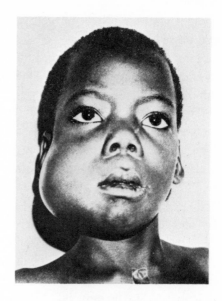

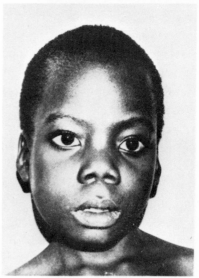

*Figure 7.5. Burkitt's Lymphoma before
and after Therapy*

H. F. OETTGEN

Other cancers do not respond so spectacularly to single chemical treatment; for example, in Hodgkin's disease a combination of several drugs in well-defined sequence has to be used.

It is reassuring, therefore, that, while the molecular biologist and the cell biologist are slowly studying control mechanisms in normal and tumor cells, others are taking a more direct approach and battle the disease with chemicals. This is fortunate because a living cell is so complex and unraveling its secrets is so slow that the empirical approaches may control cancer before we truly understand it.

Sometimes a drug is developed for the purpose of interfering with a certain reaction in the body, but it turns out to be effective for an entirely different reason. This was the case with the drug "imidazole carboxamide," which was designed to compete for bases needed for nucleic acid synthesis in the cancer cell. However, it attaches itself to nucleic acids by a chemical reaction, alters their structure, and prevents duplication of nucleic acids and of cells. If I were walking around with a 40-pound melanoma on my body making me look like some prehistoric monster, I would not care what the mechanism of action of imidazole carboxamide is. I would be grateful for it since it effects a cure in about 40 percent of the cases.

Actually, there is an excellent historic precedent for the control of a disease long before the mechanisms of the biological effects were understood. Insulin was given to the grateful diabetic patient some 50 years ago. Over the subsequent decades we decoded the pathway of metabolism of sugars, an extraordinarily complex series of reactions involving perhaps as many as 40 different enzymes which must act sequentially for the normal utilization of sugar. To date, we still do not understand completely the molecular mechanism of insulin in that sequence of reactions. We know just about everything about insulin: We know how it is synthesized in the body, the sequence of amino acids that compose it; we know its solid structure from x-ray analysis, indeed, we can synthesize it in the laboratory; but how it exerts its magic influence, of this we know nothing.

Unfortunately, the hope of a single natural agent like insulin in cancer is probably vain; diabetes is a single disease and cancer is many diseases of a hundred different cells.

Time present and time past are present in time future.

T. S. ELIOT

Chapter eight

The Next Chapter

The next chapter in the story is now being written in hundreds of laboratories of experimental biology all over the world: England, China, Russia, Hungary, and our country. The vast panorama of biological knowledge on which the curtain was lifted by American scientists with the aid of generous funding by the American people since World War II has become a goad for other societies as well. Scientists are trained, encouraged, and provisioned in all societies that can afford it. Since the demise and de-apotheosis of Stalin, the biological scientists are among the freest members of the People's Democracies. Provided they keep their political noses clean, they are permitted not only to travel to international meetings but also to spend as long as a year or more as journeymen in French, English, and

American biological laboratories. They take back with them not only new skills but new attitudes as well.

The difference in attitudes in the post-Stalin era is startling. When I visited research institutes in Hungary in 1961, the young scientists presented to me spoke either in Hungarian or in pidgin English—both equally difficult for me since I only remember the rudiments of that primitive language which is more suited for operettas than for science. (Since they are impoverished in vocabulary they create technical terms by the fusion of several simpler ones.) Unlike young scientists of the Western countries, they did not talk about their immediate experiments; instead they would recite an apparently well-rehearsed litany: "During the next three years we will demonstrate thus and so." This was amusing and revealing to someone who has never known what he might be doing three months, let alone three years ahead.

I observed a happy change there during a brief visit in 1972. Almost all scientists speak English, so one can present a seminar fluently without attempting to translate one's ideas into pidgin English. (You can always tell if a foreign audience is really with you: You slip in a simple joke and if the audience laughs spontaneously it is following you.) The young men no longer recite their three-year plans to the visitor; they describe, often in American slang, their struggles with some recalcitrant technique or problem which is their current preoccupation. Their admission of lack of progress in the presence of the head of the laboratory is revealing of both their sense of freedom and the chief's professional attitude.

The research problems they are working on are on the frontiers of biological science, and they are not required to justify their pursuit by the promise of immediate bounties in health or riches.

Meanwhile, paradoxically, in the fountain of total scientific freedom, our country, the atmosphere is changing. There is an agonizing reappraisal—agonizing to the scientist—of the interaction of science and society. Such a reappraisal was bound to come; it is not entirely the result of financial strictures produced by the bankrupting fiasco in Southeast Asia.

Careers in science and the funding of science expanded steadily after World War II. We enlarged training programs until the demand exceeded the supply, and a few individuals with marginal motivation and ability were being guaranteed careers in science. I recall an almost painful shock when I first encountered two graduate students who frankly stated that they wanted to be scientists because of the security and comfort such a career offers. They were in for a rude shock. Funding of research was expanding at a rate which, if continued, could absorb the gross national product by the year 2000.

Then the boom was lowered. Suddenly, severe cutbacks in funds for research and training resulted in utter chaos in the universities. The drain on the resources of some universities was so great they could not meet their commitments for salaries even of senior members of the staff whose support from sources outside of the university suddenly vanished. The sad truth is that ladies' hairdressers have better job security in our society than do scientists.

The probability of funding a new biomedical research program from Federal funds shrank from about 70 to 30 percent or less. Even well-established scientists await with bated breath decisions for the continued funding of their research.

At the same time, pressure for the pursuit of research relevant to immediate medical goals keeps mounting. On the surface this appears to be a valid demand: Public funds should be used for the public good. But, if there is any lesson from the history of science, it is that often it is impossible to discern the ultimate meaning of even the most esoteric or seemingly trivial finding. A little over twenty years ago a Frenchman whose worth was so little appreciated that his laboratory was relegated to the attic of the Pasteur Institute was engaged in a most recondite pursuit. He was shocking bacteria with small doses of ultraviolet irradiation. Whether his findings which I described on pages 157 to 161 had any broad biological significance was doubted even by expert experimental biologists, some of whom predicted that the phenomenon would soon be relegated to some dusty archive of esoterica.

At that time Lwoff's laboratory was supported by a small grant from our own National Institutes of Health, the French government not having yet awakened to its responsibility to French science. It was the best investment of about $10,000 a year that our government and people could have made. Those experiments were the fountainhead of the vast rivers of knowledge which contain our current understanding of virus-host relationships and much of our current knowledge of biological regulation and of cancer.

Yet, if a similar program without a clear-cut adumbration for its future potential were presented today, it would be rejected by both of the two possible sources of funding. As an abstract basic biological problem it would be examined and ripped apart by a parcel of young scientists recruited from academe who serve as the primary reviewers for such grants.

Since the emphasis is on youth for such panels the members often are people who made a quick reputation on refining past findings rather than opening a new area. And refining is all they know: The critique of the project would be that it was ''not studied in depth.'' The simple truth is that much less judgment is required to approve 70 percent of applications than to disapprove 70 percent. The latter requires both technical competence and broad vision of potential worth.

Administrators with a practical orientation would turn the project down because its relevance to mammalian systems is not apparent. Indeed, realization of the significance of Lwoff's finding was limited to the few who made pilgrimages to his laboratory. (Support for Lwoff from our National Institutes of Health was withdrawn during the Eisenhower Administration.)

If the continued productivity of many of our current scientists is uncertain, the careers of future scientists are in total jeopardy. The effectiveness of biological research in our country, whose general high quality is universally admitted, is due in large measure to the career opportunity in science for young people of modest means via the federal funding of their training. A number of studies have established that for some unknown reason American scientists come from the lower economic echelons. Training of a biological scientist is long and costly; nor is a lucrative career such as that of a successful physician or lawyer awaiting him. You cannot borrow on the prospects of such a career. Moreover, the thought of indenturing himself to the tune of $20,000 is frightening to a 22-year-old whose family has never owned that kind of money at one time. Yet the Nixon administration eliminated completely the funds for the training of scientists.

On the other hand, we are launching crusades to conquer cancer, sickle cell anemia, heart disease, and everything else that ails us. But who will be the crusaders in the laboratory twenty years from now?

It is a pity, for there is much in the American spirit—independence, rejection of authority, tinkering ability—that has kept us in the forefront of biomedical science. If present trends continue, we may lose our preeminence. It is a pity, too, for there is much left do. There is a minor school of biologists with an overly articulate apostle who claim that molecular biology is completed, it is a finished edifice; nothing new, just refinements of previous principles will emerge from its further development. This is nonsense. This was precisely what was said of physics at the turn of the century; only more refined measurements were left to be done. The brilliance of the atomic physicists of the subsequent decades illuminated the source of the defeatist resignation of classical physicists: their lack of imagination.

To fill accurately the vast lacunas in our knowlege of the intimate mechanisms of control of differentiation, and of the initiating molecular forces which convert a potentially normal cell to malignancy, will yet take perhaps a century, or more, of dedicated, inspired effort by hundreds of molecular and cell biologists. Splendors invented by the genius of Evolution equal to those already unveiled are awaiting the questing minds and refined tools of biologists yet unborn.

What is needed is a stable, dependable level of support of both current biological research and the training of our best young people to roam the frontiers of biology decades from now. In the past few years such support appeared to be dictated by a roulette wheel rather than any guiding policy. Our federal funding of science is not unlike the finances of an old-fashioned family grocery store. You never knew from month to month what was available; at the end of a good year enough was accumulated for new shoes all around or even for a fur coat; in bad years there was nothing.

As of this moment, because of an impasse between the President and Congress, the National Institutes of Health, the main agency for funding health-related research, has been without a real budget for the whole year. Hundreds of projects are being disbanded for want of funding. The effect of this is harrowing to the workers in the field and disastrous for science.

It is curious how readily a commitment of a score of billions of dollars was made to take our astronauts to the Moon, but how quixotic is commitment to the continuing exploration of our own biological world. The potential gain from the latter is infinitely greater than from the former. To be sure, as an adventure of the human spirit and mind nothing can excel our safe excape from our gravitational shackles, but the increment of scientific information—in spite of the spectacular claims made by the National Space Agency—which accrued from the many trips to the Moon is miniscule. We already knew infintely more about the surface of the moon from exploration by earth-bound instruments than we know about regulatory mechanisms in a single cell in our little finger.

The decoding of secrets, as yet hidden in our cells, has inestimable potential value. What is the value of the discovery of insulin which has enriched and lengthened the lives of millions during the past fifty years? It is the firm conviction of many of us on the frontiers of biology that biological agents and mechanisms of equal worth and potency are awaiting discovery.

Among the latter we count the mechanism of brain function, which is admittedly beyond the reach of our current molecular and cell biology. Will we ever penetrate its mystery? Our leading Cassandra of biology rejects this challenge as being totally beyond our reach: Evolution has not endowed our mind with sufficient power and finesse to penetrate this last arcanum. To be sure, progress is almost nil in this area. But that is no reason for a supine admission of impotence in the face of this ultimate barrier.

In the absence of knowledge one is prone to take refuge in mysticism. Until sentient man appeared, the splendors of Evolution were lost for want of an appreciative witness. Could a saber-toothed tiger value the triumph of the myriads of sequential mutations which produced it? Will the finest triumph of Evolution, the molecular mechanism of brain function, remain hidden and be forever unappreciated? One thinks not. Evolution, it seems, strove to create the uniquely powerful human mind to be appreciated at long last. Unlike all other creations of Evolution there was no gradualism in the emergence of the human brain. All other brains are puny and primitive compared to ours. If there be vanity in Nature she will not keep us at arms length from solving this last mystery and admiring her ultimate invention.